Advanced Level Biology

AS Biology with Stafford

Unit One: Lifestyle, Genes & Health

Paper reference: 6BIO1

Typeset & layouts by   : Mohamed Sobir
Cover designed by   : Mohamed Sobir
Printed in Maldives by   : Copier Repair
Published by   : Author publisher

ISBN: 978-81-910705-0-7

Contact details of the author:

Stafford Valentine Redden
Near Hindhustan Oil Mill,
Ramchandrapur, Jatni P.O,
Orissa, India, 752050
staffordv@yahoo.com
+960 7765507

Dedicated to my loving Mum & Dad

Contents

Chapter One
Water – The Solvent for Life 1

Chapter Two
Carbohydrates 4

Chapter Three
Lipids 10

Chapter Four
The Circulatory System 12

Chapter Five
Cardio-Vascular Diseases 22

Chapter Six
Risk Factors 25

Chapter Seven
Treatment of Cardio-Vascular Diseases 30

Chapter Eight
Cholesterol & Cardio- Vascular Diseases 35

Chapter Nine
Obesity 38

Chapter Ten
Correlation, Causation & Study Design 45

Chapter Eleven
Cell Membrane Structure 54

Chapter Twelve
Transport Across Membranes 58

Chapter Thirteen
Gas Exchange in Living Organisms 64

Chapter Fourteen
Structure of Proteins 67

Chapter Fifteen
Enzymes 71

Chapter Sixteen
Nucleic Acids 76

Chapter Seventeen
DNA Replication 79

Chapter Eighteen
The Genetic Code & Protein Synthesis 82

Chapter Nineteen
Mutations 87

Chapter Twenty
Monohybrid & Inheritance 89

Chapter Twenty One
Cystic Fibrosis 93

Chapter Twenty Two
Gene Therapy 96

Topic Two
Genes & Health

Topic One
Lifestyle, Health & Risk

CHAPTER ONE
WATER – THE SOLVENT FOR LIFE

Water is a polar covalent compound. Due to difference in electro-negativities of oxygen and hydrogen, one end of the water molecule bears a slight negative charge, while the other end bears a slight positive charge, as shown in fig.1.1. This is called the dipolar nature of water.

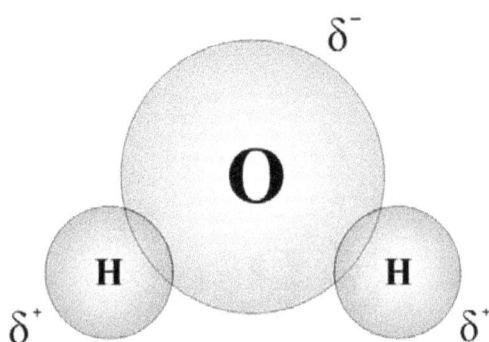

Fig.1.1

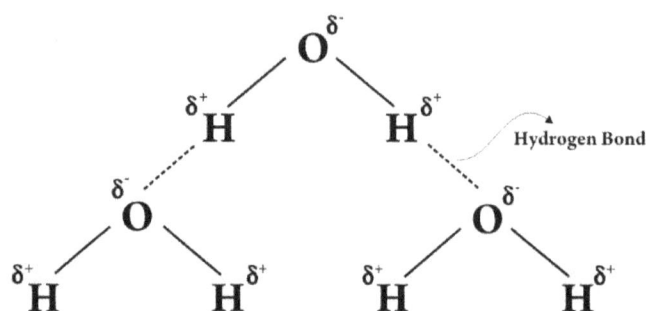

Fig.1.2

The positive end of one water molecule is attracted to the negative end of another water molecule. This force of attraction is called a hydrogen bond, shown in fig.1.2. The dipolar nature of water is responsible for the following properties of water.

The polar nature of water molecule makes it a good solvent (Universal solvent)
Water is the universal solvent for life. The partial charge that develops across the water molecule helps make it an excellent solvent. Water dissolves many substances by surrounding charged particles and "pulling" them into solution, as shown in fig.1.3. For example, common table salt, sodium chloride, is an ionic substance that contains alternating sodium and chloride ions.

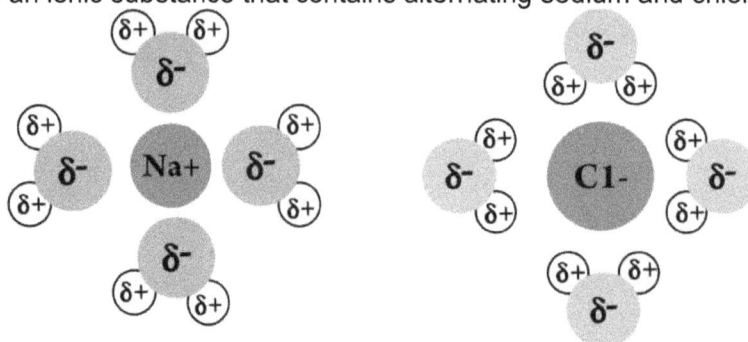

Fig.1.3

When sodium chloride is added to water, the partial charges on the water molecule are attracted to the Na^+ and Cl^- ions. The water molecules work their way into the crystal structure and between the individual ions, surrounding them and slowly dissolving the salt. The negative oxygen ends of water molecules will surround the positive sodium ions; the positive hydrogen ends will surround the negative chloride ions, as shown in fig.1.3. This idea also explains why some substances do not dissolve in water. Oil, for example, is a non polar molecule. Because there is no net electrical charge across an oil molecule, it is not attracted to water molecules and therefore does not dissolve in water. Any molecule with a polar region will dissolve. Water molecules gather around large protein molecules to form a special kind of solution called a **colloid**, as found in the cytoplasm of cells.

Almost all ionic compounds and small organic compounds are soluble in water. This enables easy transport of materials. **Sap is transported** through xylem and phloem. The sap consists of sucrose, amino acids and ions **dissolved in water**. Glucose, amino acids, hormones, urea and many other substances **dissolve in the blood plasma** and can be easily **transported in the blood**. Sperms and ova in mammals are also transported in watery fluids. Water also provides a good medium for soluble substances to collide with each other and react. These collisions would not be possible or would be too slow in solid state. Water also plays an important role in the maintenance of the tertiary structure of enzymes and the phospholipid bilayer of the cell surface membrane. Water is very cohesive. This allows water to be pulled along a pathway with relative ease. Water is also a good adhesive. It will cling on to many objects and act as a glue. Capillary Action is an example of **cohesion and adhesion** working together to move water up a thin tube like the xylem in plants. Imbibition is the process of soaking into a hydrophilic substance. Water being taken into a sponge, into a seed or into paper towels are examples of Imbibition. Transparency allows the transmission of light e.g. for photosynthesis in aquatic habitats

Water has a high latent heat of vaporization: The latent heat of vapourisation is the amount of heat energy needed to vapourise 1 kilogram of water at a constant temperature. In humans (at 37 ^{0}C), this value is 539 000 cal/Kg. So, when water evaporates from a surface it absorbs latent heat of vapourisation from the surface and cools the surface down. This makes water a good coolant. This is especially useful in cooling of plant tissues by transpiration and cooling of mammals by sweating or panting.

Water has a high latent heat of Fusion: The latent heat of fusion is the amount of heat energy needed to convert 1 kilogram of water into ice at a constant temperature. This value for water is about 80 KJ/g. The high latent heat of fusion ensures that water bodies do not freeze easily in cold climates. It also prevents the cytoplasm from freezing at low temperatures.

Water has a high specific heat capacity – The specific heat capacity is the amount of heat energy needed to increase the temperature of 1 kilogram of water by 1^{0}C. The value for water is 4200 J/Kg ^{0}C. So, a lot of heat must be added or removed to change the temperature of water. This property of water prevents sudden fluctuations in temperature of organisms or aquatic environments, allowing enzymes to function at their optimum temperature. This high specific heat allows water to act as a heat sink. Water will retain its temperature after absorbing large amounts of heat, and retain its temperature after losing equally large amounts of heat. The reason for this is that Hydrogen bonds must absorb heat to break. They must release heat when they form. The Ocean acts as a tremendous heat sink to moderate the earth's temperature.

Surface tension is the property of a liquid which makes its surface behave like a stretched membrane, mainly caused due to hydrogen bonding between water molecules, as shown in fig.1.4. This allows materials to rest upon it if the surface tension is not broken. Pollen, dust, water insects, and other biological materials are able to remain on the surface of the water because of this tension. Mosquito larvae also use the surface tension of water to cling to the surface and breathe air, through siphons. Surface tension decreases the ease with which gases dissolve into water.

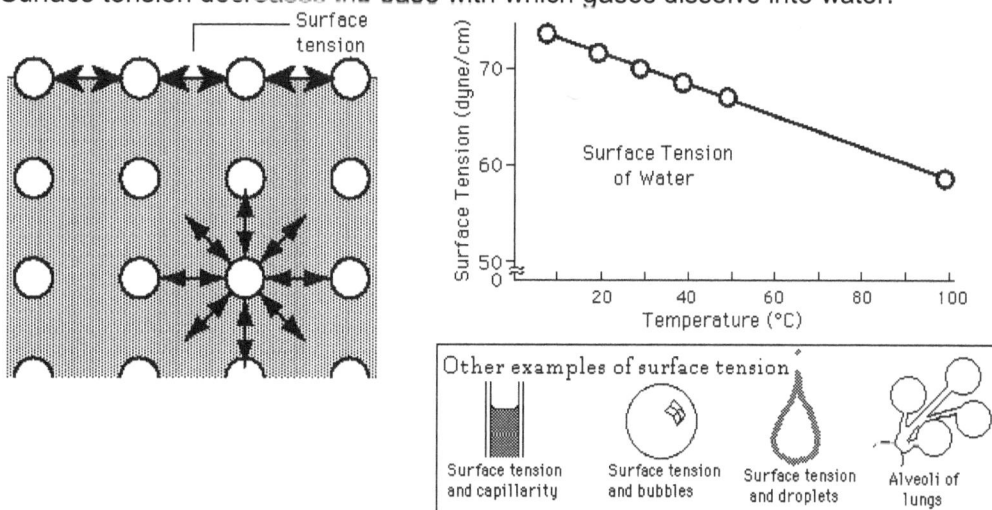

Fig.1.4

Water has maximum density at 4^0 C. This means that the densest water at 4^0 C will remain at the bottom of an aquatic habitat. This prevents aquatic habitats from freezing completely, so that aquatic organisms can survive at the bottom, unfrozen at 4^0 C. Ice floats because maximum Hydrogen bonding occurs at 0^0 C. The ice on the surface of a frozen lake will serve as **an insulating layer** and reduce heat loss from the water body, so that the rest of the water at the bottom does not freeze.

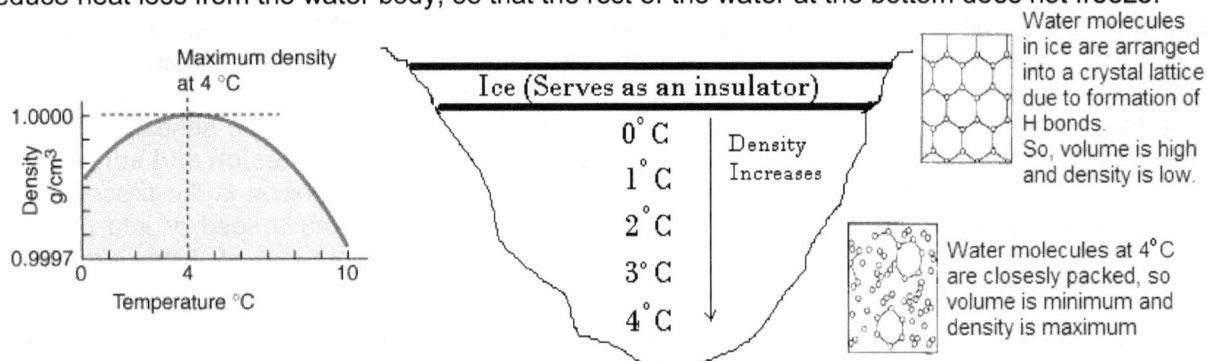

Maximum density at 4 °C

Density g/cm³

1.0000

0.9997

0   4   10

Temperature °C

Ice (Serves as an insulator)

0° C
1° C
2° C
3° C
4° C

Density Increases

Water molecules in ice are arranged into a crystal lattice due to formation of H bonds. So, volume is high and density is low.

Water molecules at 4°C are closesly packed, so volume is minimum and density is maximum

Fig.1.5

CHAPTER TWO
CARBOHYDRATES

Learning outcomes: by the end of this chapter you should be able to

Edexcel Syllabus Spec 3: Distinguish between monosaccharides, disaccharides and polysaccharides (glycogen and starch — amylose and amylopectin) and relate their structures to their roles in providing and storing energy (β-glucose and cellulose are not required in this topic).

Edexcel Syllabus Spec 4: Describe how monosaccharides join to form disaccharides (sucrose, lactose and maltose) and polysaccharides (glycogen and amylose) through condensation reactions forming glycosidic bonds, and how these can be split through hydrolysis reactions.

Carbohydrates are polyhydroxy aldehydes or ketones, containing Carbon, Hydrogen and Oxygen. They have the general formula $(CH_2O)_n$. A few carbohydrates are shown in fig.2.1.

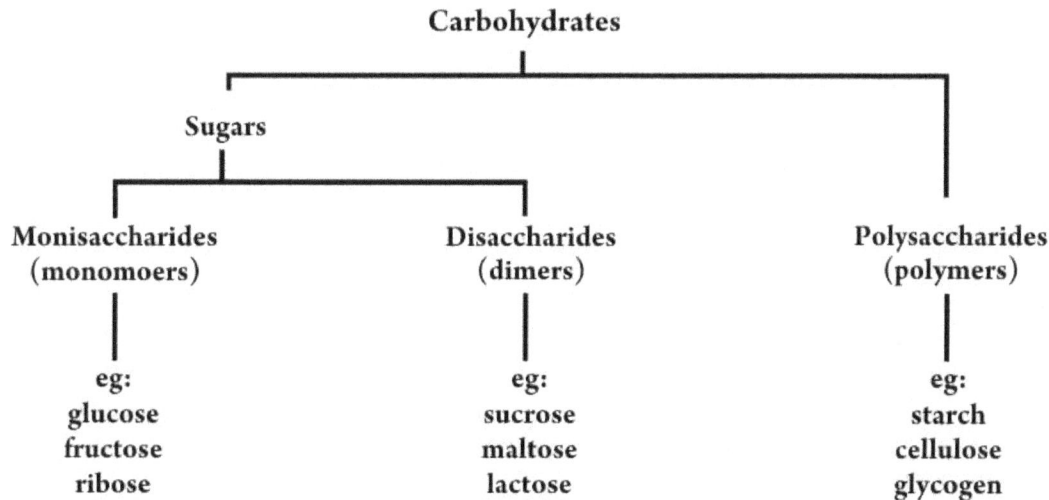

Carbohydrates

Sugars

Monisaccharides (monomoers)	Disaccharides (dimers)	Polysaccharides (polymers)
eg: glucose fructose ribose	eg: sucrose maltose lactose	eg: starch cellulose glycogen

Fig.2.1

Monosaccharides - monomers

The simplest carbohydrates are called monosaccharides. The general formula of a monosaccharide is $(CH_2O)_n$, where n is the number of carbon atoms in the molecule. A six-carbon monosaccharide is called Hexose and a five-carbon monosaccharide is called a pentose sugar. Examples of some monosaccharides are shown below.

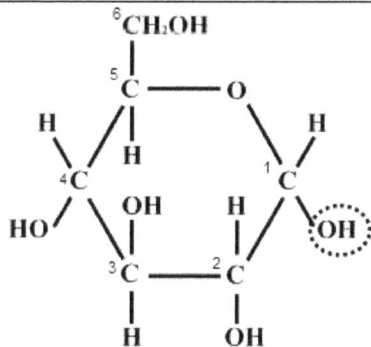

Fig.2.2

Alpha glucose ($C_6H_{12}O_6$) is a component of most disaccharides - Maltose, sucrose and lactose; and many polysaccharides - Glycogen, amylose and amylopectin. It is also the primary substrate in cellular respiration and provides the energy for most metabolic reactions. Carbohydrates are transported in mammalian blood as glucose.

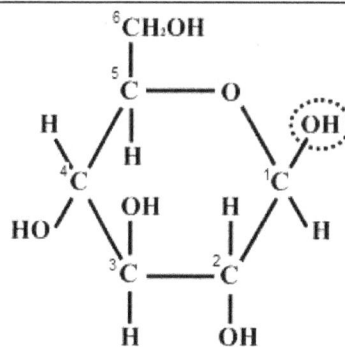

Fig.2.3

Beta glucose ($C_6H_{12}O_6$) is a component of the cellulose micro fibrils that are found in the cell walls of plants. Note the position of OH on the first Carbon atom of glucose.

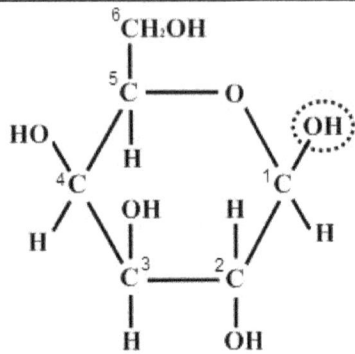

Fig.2.4

Galactose **(C_6H_12O_6)** is another monosaccharide, which is a component of lactose sugar, found in mammalian milk.

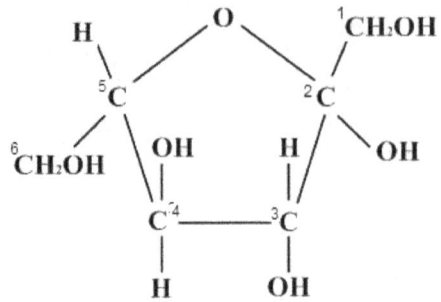

Fig.2.5

Fructose (C_6H_12O_6) is a component of sucrose. It is commonly found in honey and fruits. It is the sweetest sugar. Twice as sweet as sucrose.

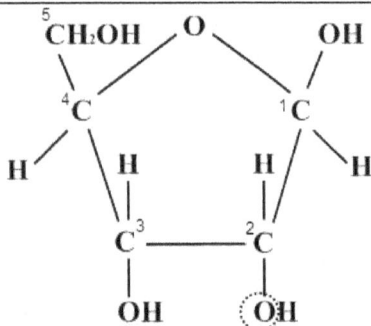

Fig.2.6

Ribose sugar (C_5H_10O_5) is found in nucleotides that form ribonucleic acid (mRNA, tRNA and rRNA). It is also a component of ATP (Adenosine Triphosphate) and NAD (Nicotine amide Adenine Dinucleotide).

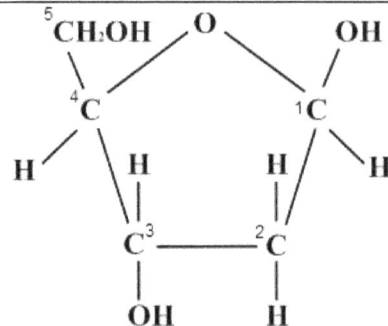

Fig.2.7

Deoxyribose sugar (C_5H_10O_4) is found in Deoxyribonucleic acid (DNA).

The fig.2.8 shows the relationship between the different forms of glucose

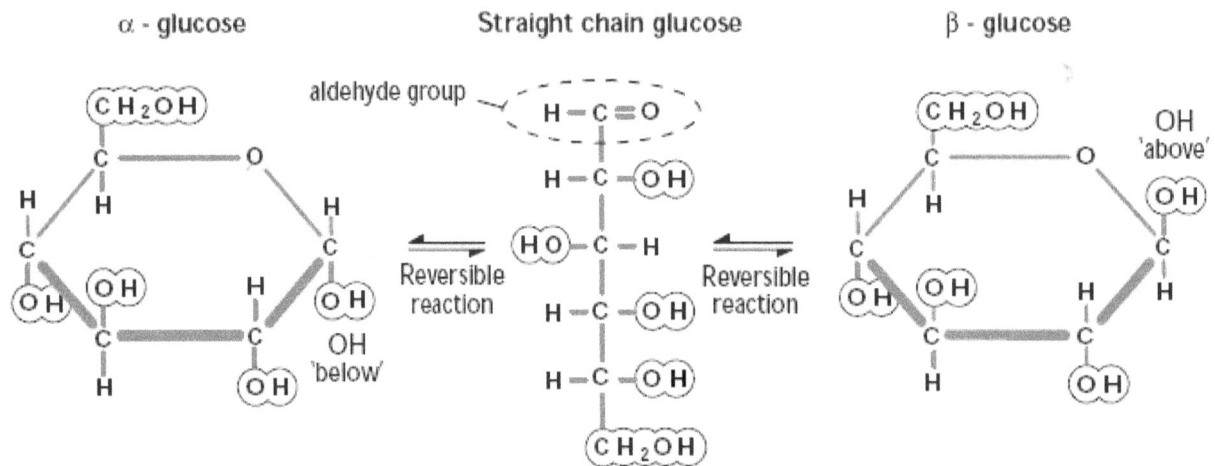

Fig.2.8

Disaccharides - dimers

Two monosaccharides can be linked by glycosidic bonds to form disaccharides. The general formula of a disaccharide is usually $2[(CH_2O)_n] - H_2O$, where n is the number of carbon atoms in the molecule. The fig.2.9 shows the structure of maltose ($C_{12}H_{22}O_{11}$), formed by the combination of two glucose residues.

Condensation reaction

During a condensation reaction, two monosaccharides are joined by the **removal of a water molecule** to form a **glycosidic bond**, as shown in fig.2.9.

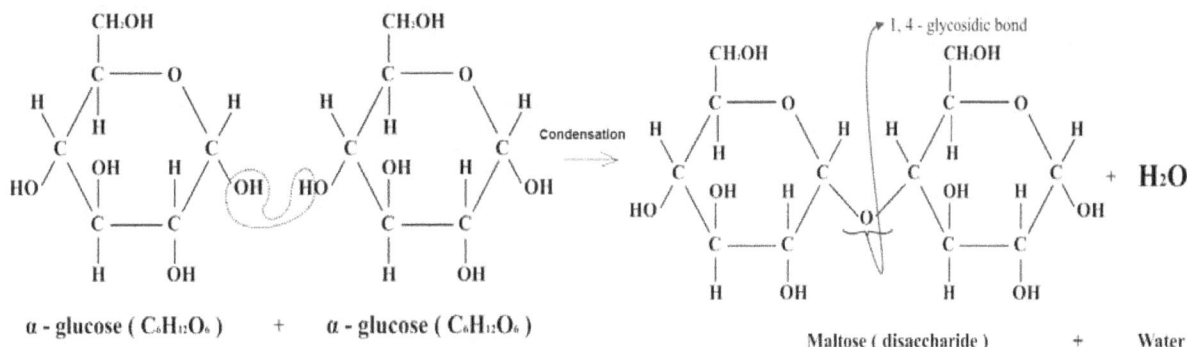

α - glucose ($C_6H_{12}O_6$) + α - glucose ($C_6H_{12}O_6$)   Maltose (disaccharide) + Water

Fig.2.9

Hydrolysis

Hydrolysis is the splitting of a disaccharide in to its monosaccharides, by the **addition of water molecules**. The breaking of glycosidic bonds by the addition of water molecules brings about hydrolysis, as shown in fig.2.10. This can be brought about by enzymes or by treating with hydrochloric acid, followed by heating.

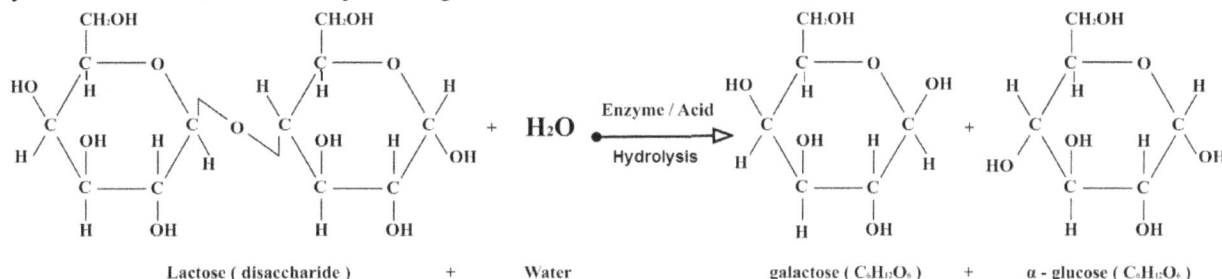

Lactose (disaccharide) + Water   galactose ($C_6H_{12}O_6$) + α - glucose ($C_6H_{12}O_6$)

Fig.2.10

Structures and roles of disaccharides
Lactose

Lactose is commonly called milk sugar. It consists of a β galactose residue linked to an α glucose residue by a 1, 4 – glycosidic bond. Lactose can be split into galactose and glucose by using the enzyme Lactase. This is a hydrolase enzyme, which hydrolyses the glycosidic bond.

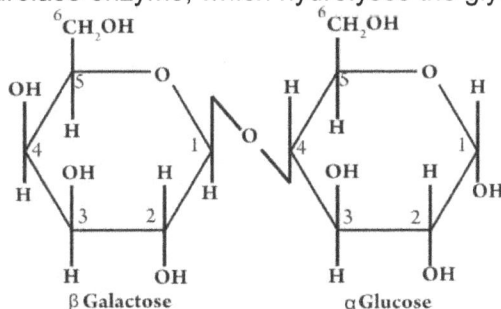

β Galactose   α Glucose

Fig.2.11

Roles: Lactose is the sugar found in mammalian milk. It serves as a carbohydrate, which provides energy to young mammals. **Galactosaemia (Galactose in the blood)** is an inherited disorder that produces an inability to metabolise galactose due the absence of an enzyme called galactose transferase. Babies with galactosaemia vomit soon after they have started breast feed because of the production of metabolic toxins. If it is not treated it can result in mental retardation. The babies obtain the galactose because it is a sub-unit of lactose which is present in milk. This can be overcome by consuming soya milk, which does not contain lactose.

Secondary lactose intolerance may arise due to the inability to produce sufficient lactase. The lactose remains undigested and cannot be absorbed by the ileum. This results in nausea, cramps, bloating of abdomen and diarrhoea. This can be treated by consuming milk treated with **immobilised** lactase. Yoghurt and ice cream are two milk products in which the lactose has been converted into lactic acid, glucose and galactose.

Maltose

Maltose is commonly called malt sugar. It is an essential ingredient in beer production. It consists of two α glucose residues linked by a 1, 4 – glycosidic bond. It can be hydrolysed into glucose by the enzyme maltase.

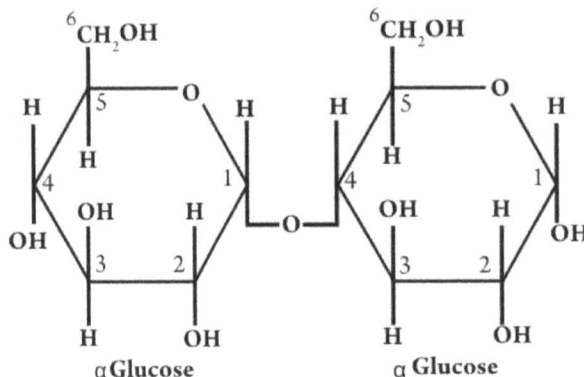

Fig.2.12

Roles: Maltose is an intermediate product of starch hydrolysis in plants and germinating seeds. It can be hydrolysed to form glucose, which is a respiratory substrate.

Sucrose

Sucrose is the common sugar that we use in puddings, tea and coffee. It consists of an α glucose residue linked to fructose by a 1, 2 – glycosidic bond. Sucrose can be split into fructose and glucose by the enzyme sucrase or invertase.

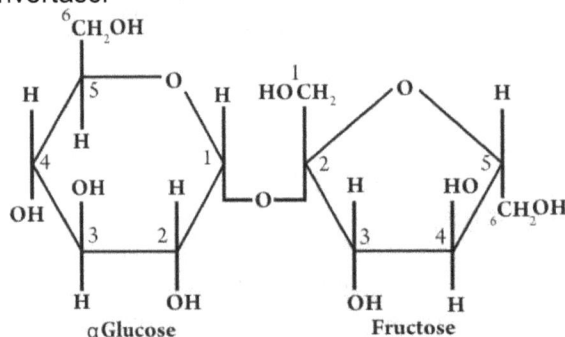

Fig.2.13

Roles: Sucrose is the form in which plants transport carbohydrates, through the phloem. Sucrose is also stored in the stems and roots of sugarcane and sugar beet plants.

Additional information

Reducing sugars will react with Benedict's solution and reduce the blue coloured Cu^{2+} ions into red colour Cu^+ ions. The sugar solution is added to Benedict's solution in a test tube and boiled in a water bath. The colour changes from blue to green to yellow and finally to red, as the concentration of the red Cu^+ ions increase. Solutions with low concentrations of reducing sugars will turn green or yellow.

Non-reducing sugars will not change the colour of Benedict's solution. However, after acid or enzyme hydrolysis, it will change the colour of Benedict's solutions.

Acid hydrolysis: Add hydrochloric acid to the sugar solution and heat it gently. Neutralise the mixture by adding Sodium Hydrogen carbonate solution or sodium hydroxide solution. Carry out the Benedict's test as explained above.

Enzyme hydrolysis: The sugar solution must be treated with enzymes and then the Benedict's test will give a positive result. Sucrose is an example of a non reducing sugar.

Polysaccharides - Polymers
Many monosaccharides can be linked by glycosidic bonds to form polysaccharides.

Starch
Starch is a polysaccharide made up of many α glucose residues linked by glycosidic bonds. Starch is a **mixture** of **amylose** and **amylopectin**. The main function of starch is that it is an **energy storage molecule** in plant cells.
Structure of starch related to its function;
- **Starch is Compact -** so it takes up less space in the cell.
- **Starch is Insoluble** – so it cannot leave the cell easily.
- **Starch is Insoluble** – so it does not have an osmotic effect.
- **Starch is Insoluble and unreactive** – so it does not get involved in chemical reactions in cell.
- Amylose and amylopectin can easily be **hydrolyzed by enzymes** into maltose.

Amylose structure
Starch is a mixture of **amylose** and amylopectin. Amylose is composed of α glucose residues linked by 1, 4 – glycosidic bonds, formed by condensation reactions. The unbranched chain is then coiled into an amylose helix due to formation of hydrogen bonds between the glucose residues. There are six glucose residues per turn of the helix. The amylose reacts with iodine to form a blue black complex, which is used as a common test for starch. Amylose can be hydrolysed by the enzyme amylase, to yield maltose.

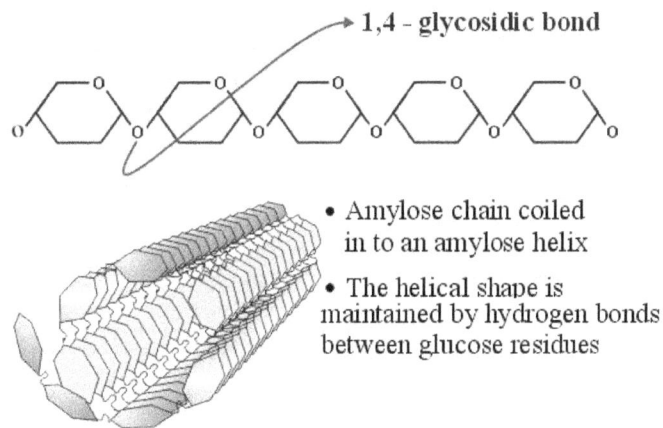

- Amylose chain coiled in to an amylose helix
- The helical shape is maintained by hydrogen bonds between glucose residues

Fig.2.14

Amylopectin structure
Amylopectin is also made up of α glucose residues. However, the glucose residues form a branched chain. The branches are formed due to the presence of 1, 6 – glycosidic bonds. The 1, 4 – glycosidic bonds are found in the unbranched part of the chain. Branches usually form after every 20 to 30 glucose residues. Hydrolysis of starch is brought about by amylase or by acid hydrolysis. Slow release starch is found in potatoes and seeds. The starch granules are trapped inside the cells and cannot easily come in contact with enzymes. In cooked tissues, the enzymes and water can easily penetrate the cells and digest the starch. The digestibility of starch may also be influenced by the relative proportion of amylose and amylopectin.

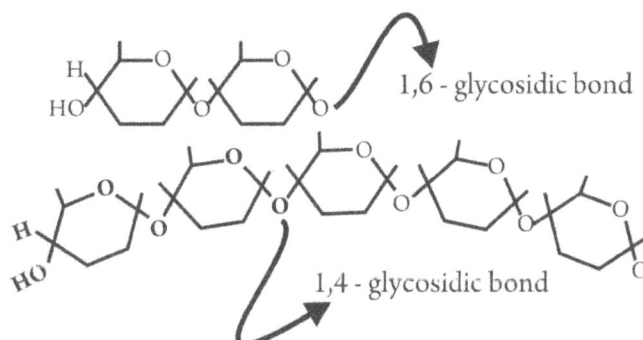

Fig.2.15

Glycogen

Glycogen is also made up of α glucose residues. The glucose residues form a branched chain. The branches are formed due to the presence of 1, 6 – glycosidic bonds. The 1, 4 – glycosidic bonds are found in the unbranched part of the chain. Branches usually form after every 8 to 10 glucose residues. Glycogen synthase is an enzyme involved in the formation of glycogen, by condensation reactions. Glycogen phosphorylase is the primary enzyme involved in the hydrolysis of glycogen. The glycogen is **stored** in the liver and muscle tissues of animals and can readily be hydrolysed to provide glucose for cellular respiration. The structure of glycogen and amylopectin are similar, but amylopectin forms branches after every 20 to 30 residues.

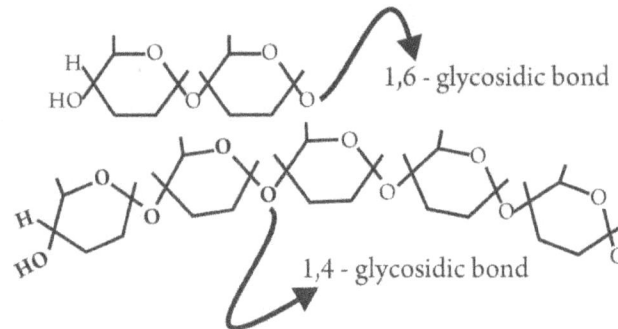

1,6 - glycosidic bond

1,4 - glycosidic bond

Fig.2.16

Structure related to function: Glycogen is **Compact** so it takes up less space in the cell. It is **Insoluble** so it cannot leave the cell easily and does not have an osmotic effect. It is relatively unreactive so it does not get involved in chemical reactions in cell. It can easily be **hydrolyzed by enzymes** into glucose and used for respiration in cells.

CHAPTER THREE
LIPIDS

Lipids are a large group of organic compounds made up of Carbon, Hydrogen and Oxygen. They are **polyesters** formed by linking of **glycerol** with **fatty acid chains** by **condensation reactions**. Fats and oils are chemically similar, but fats are solid and oils are liquid at room temperature. Waxes have long chained alcohols linked to their fatty acids. All lipids are **non polar**, hence **insoluble in water.**

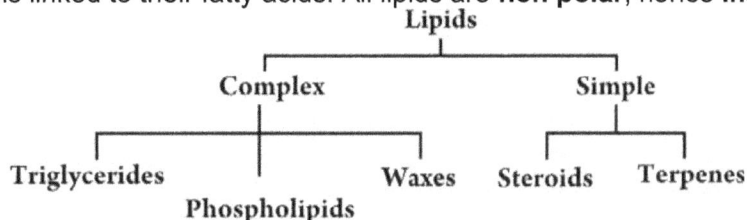

```
                        Lipids
              ┌───────────┴───────────┐
           Complex                  Simple
        ┌─────┴─────┐          ┌──────┼──────┐
  Triglycerides   Waxes    Steroids   Terpenes
     Phospholipids
```

Triglycerides are the most common form of lipids. It consists of one glycerol molecule and three fatty acid chains. The hydoxyl (OH) groups of glycerol and the carboxyllic acid (-COOH) group of each fatty acid are linked by an ester bond formed by condensation reactions, as shown in fig.3.1.

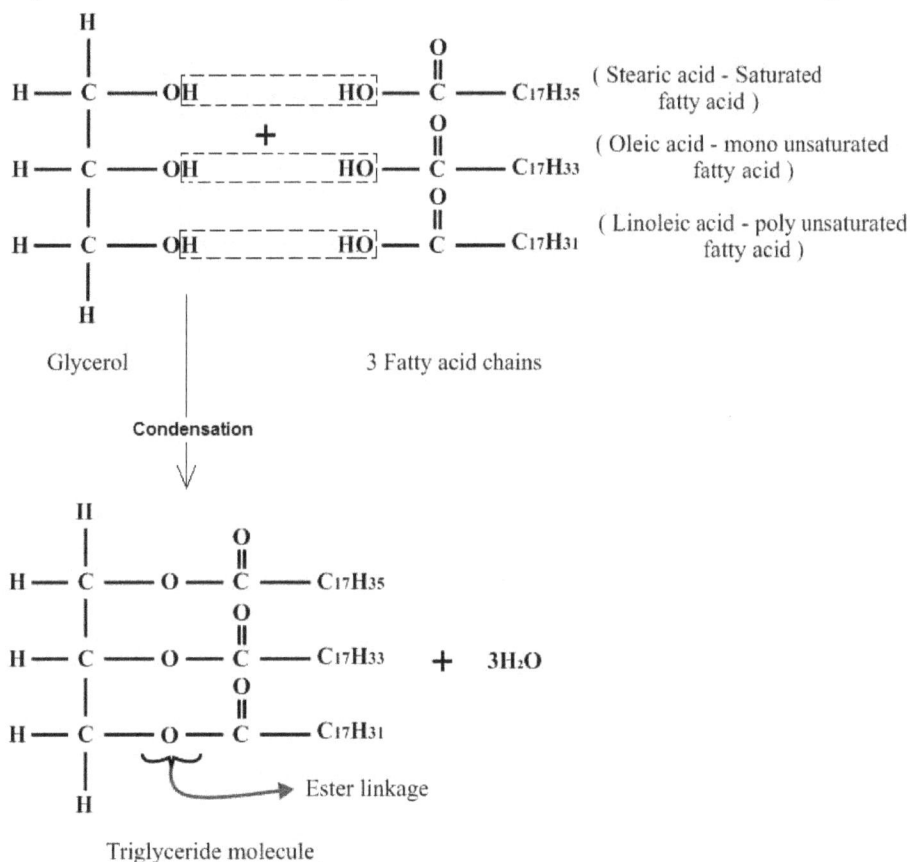

Triglyceride molecule

Fig.3.1

Saturated fatty acids have only single bonds between carbon atoms (C-C). These fatty acids form straight chains, and have a high melting point. Hence, they are usually solids at room temperature. They have the general formula $C_nH_{2n+1}COOH$. Fats such as butter have a higher proportion of saturated fatty acids. Palm oil and coconut oil also have a high concentration of saturated fatty acids.
Example: Stearic acid ($CH_3(CH_2)_{16}COOH$)
Unsaturated fatty acids have at least one double bond (- C = C -) **between carbon atoms** in the

hydrocarbon chain. These fatty acids form **bent** chains, and so have a low melting point. Hence they are usually liquids at room temperature. Fatty acids with one double bond are called Monounsaturated fatty acids (MUFAs). Fatty acids with more than one double bond are called polyunsaturated fatty acids (PUFAs). Unsaturated fatty acids have lesser number of Hydrogen atoms than a saturated hydrocarbon with the same number of Carbon atoms. **Examples:** oleic acid – MUFA, linoleic acid - PUFA.

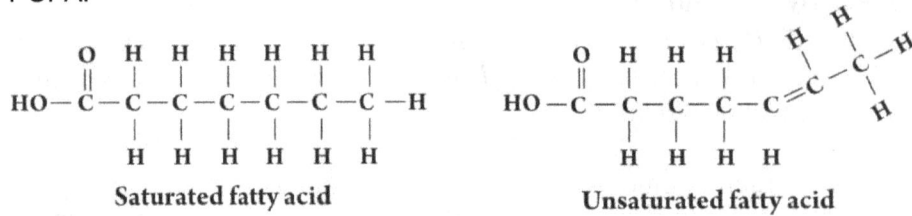

Saturated fatty acid   Unsaturated fatty acid

Fig.3.2

Key points

- Lipids include fats, oils, waxes, phospholipids and steroids.
- Triglycerides are composed of three (tri) fatty acids combined by condensation reactions to form ester bonds with glycerol.
- Fatty acids have long hydrocarbon chains and a carboxyl group (COOH).
- Saturated fatty acids (and hence saturated fats) have no double bonds in the hydrocarbon chain which is 'saturated' with hydrogen.
- Unsaturated fatty acids have one double bond between the carbon atoms.

Additional material – will be useful for topic two

__Phospholipids__ are another class of lipids. The diagram below shows a molecule of a phospholipid. It has two fatty acid chains and one phosphate group joined to a glycerol molecule by condensation reactions. The phosphate side of the molecule is polar and the fatty acid region is non polar. The cell membranes are composed of phospholipids, which form a bilayer.

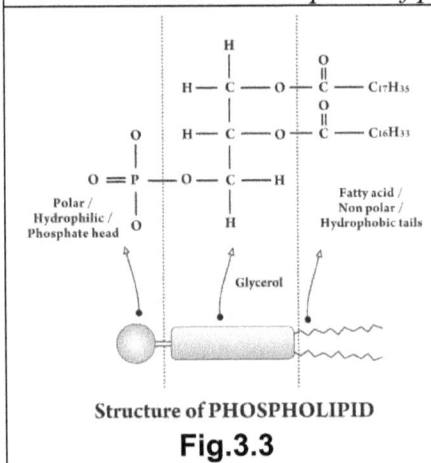

Structure of PHOSPHOLIPID
Fig.3.3

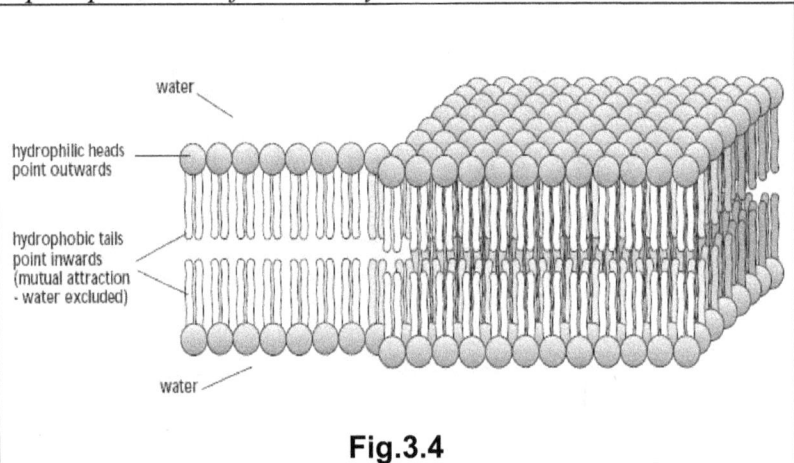

Fig.3.4

__Cholesterol__ is another important lipid component or cell membranes although it is very different in chemical structure to the lipids so far described. It has a ring structure and belongs to a group called steroids which include the sex hormones testosterone and oestrogen. Cholesterol molecules are strongly hydrophobic and they take refuge between the tails of phospholipids in membranes.

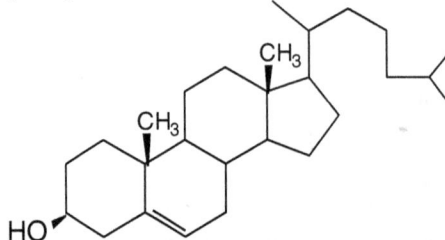

Fig.3.5

CHAPTER FOUR
THE CIRCULATORY SYSTEM

Learning outcomes: by the end of this chapter you should be able to

Edexcel Syllabus Spec 6: Explain why many animals have a heart and circulation (mass transport to overcome limitations of diffusion in meeting the requirements of organisms).

Edexcel Syllabus Spec 7: Describe the cardiac cycle (atrial systole, ventricular systole and diastole) and relate the structure and operation of the mammalian heart to its function, including the major blood vessels.

Edexcel Syllabus Spec 8: Explain how the structures of blood vessels (capillaries, arteries and veins) relate to their functions.

The need for a circulatory system

One of the main functions of the circulatory system in mammals is the transport of respiratory gases, metabolites, metabolic wastes, and hormones. As organisms get bigger their surface area / volume ratio gets smaller, as shown in fig.4.1. So, as organisms become bigger it is more difficult for them to exchange materials with their surroundings.

The single celled organisms exchange food, respiratory gases and excretory products from the environment, directly from the cell surface membrane. However, animals are multicellular and have a very **low surface area to volume ratio** and **high diffusion distance** from the surface to the body core. **Diffusion from the body surface would be too slow to supply or remove materials from the cells at a suitable rate to sustain metabolism in these cells**. So, multicellular organisms need a circulatory system to supply or remove materials from the cells at a rapid rate.

ORGANISM	LENGTH	SA (M^2)	VOL. (M^3)	S/A:VOL
Bacterium	1 μm	6 x 10^{-12}	10^{-18}	6,000,000:1
Amoeba	100 μm	6 x 10^{-8}	10^{-12}	60,000:1
Fly	10 mm	6 x 10^{-4}	10^{-6}	600:1
Dog	1 m	6 x 10^{0}	10^{0}	6:1
Whale	100 m	6 x 10^{4}	10^{6}	0.06:1

Fig 4.1

The cubes in fig.4.2 show the relationship between size and surface area to volume ratio.

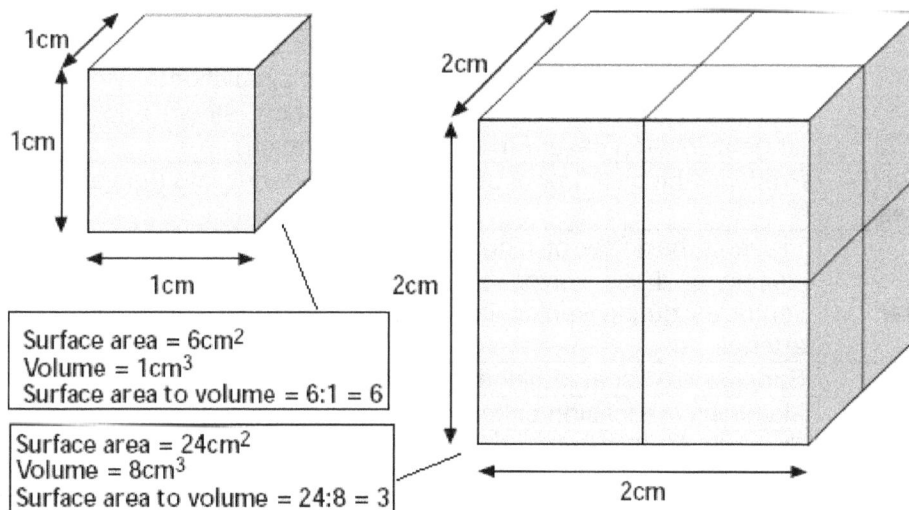

1cm
1cm
1cm

Surface area = 6cm^2
Volume = 1cm^3
Surface area to volume = 6:1 = 6

Surface area = 24cm^2
Volume = 8cm^3
Surface area to volume = 24:8 = 3

2cm
2cm
2cm

Fig 4.2

Structure of mammalian heart

The mammalian heart is located in the thoracic cavity. It is embraced by the lungs and slightly tilted to the left. That is why the powerful contraction of the left atrium can be felt on the left of the chest. The heart is enclosed in membranes called the pericardial membranes, which contain the pericardial fluid between them. The fluid absorbs mechanical shock.

The wall of the heart is made up three distinct layers:

- The outer epicardium – consisting of flattened epithelial cells and supporting connective tissue;
- a thick layer of myocardium – consisting of cardiac muscles cells; and
- an inner endocardium – made up of flattened epithelial cells and connective tissue.

Fig.4.3 shows the major structures in the heart.

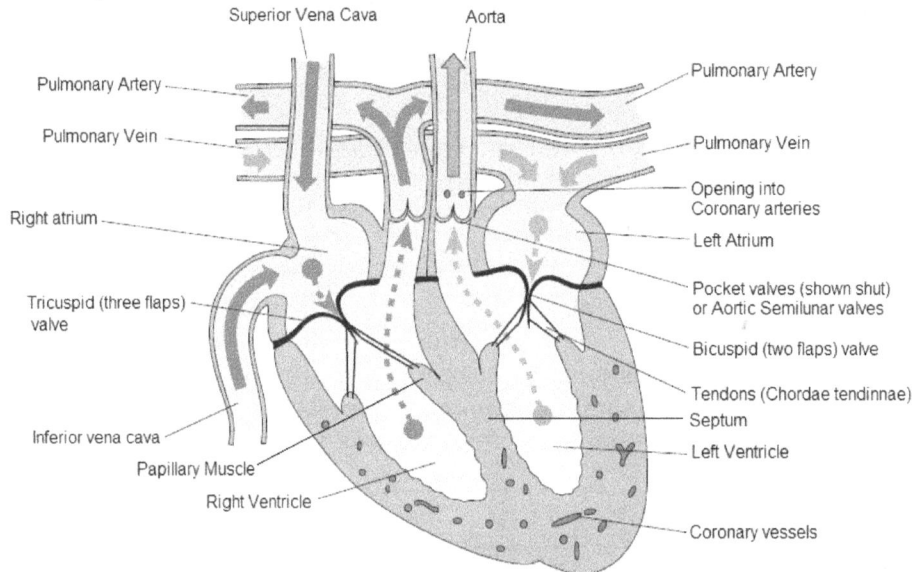

Fig.4.3

Structure	Function
Superior Vena Cava	Carries deoxygenated blood from the **head** to the **right atrium**.
Inferior Vena Cava	Carries deoxygenated blood from the **body** to the **right atrium**.
Aorta	Carries oxygenated blood from the **left ventricle** to the **body**.
Pulmonary artery	Carries deoxygenated blood from the **right ventricle** to the **lungs**.
Pulmonary vein	Carries oxygenated blood from the **lungs** to the **left atrium**.
Left Atrium	Receives oxygenated blood from pulmonary vein and pumps it into the left ventricle
Right Atrium	Receives deoxygenated blood from vena cava and pumps it into the right ventricle
Left Ventricle	Receives oxygenated blood from the left atrium and pumps it to all the organ systems at high pressure (systemic circulation)
Right Ventricle	Receives deoxygenated blood from the right atrium and pumps it to the lungs at lower pressure (pulmonary circulation)
Pocket valves or semilunar valves	Prevents the backflow of blood from the arteries into the ventricles.
Bicuspid and tricuspid valve (atrio-ventricular valves)	Prevents backflow of blood into the atria during ventricular systole. The closing of these valves also helps to build up a high pressure in the ventricles during ventricular systole, which helps to push blood into the arteries.
Coronary artery	Supplies oxygenated blood to the cardiac muscle
Papillary muscles	Maintains the tension on the *chordae tendinnae*
***Chordae tendinnae* (tendons)**	Prevents the bicuspid and tricuspid valves from flipping over into the atria during ventricular systole.
Septum	Separates the right and left sides of the heart, so that oxygenated blood in the left cannot mix with deoxygenated blood in the right side of the heart.

The walls of the **left ventricles are much thicker** as they have to exert a greater pressure to pump blood to all parts of the body, even to the head against the force of gravity. The right ventricles exert a much lower pressure as the lungs are close to the heart and pulmonary tissues are very delicate. High pressure in the pulmonary circulation will lead to **pulmonary oedema**, causing the lungs to fill with tissue fluid. This will cause the person to drown in their own tissue fluid. The left side of the heart contains only oxygenated blood and the right side contains only deoxygenated blood. The right and left sides are separated by a muscular wall called as the **septum**.

Double circulatory system
The mammalian circulatory system consists of two circulations, as shown in fig4.4
Systemic circulation is the pumping of oxygenated blood to all the organ system of the body. It is a *high pressure circulation* originating from the left ventricle and terminating at the right atrium, which receives deoxygenated blood from the organ systems or tissues.
Pulmonary circulation is the pumping of deoxygenated blood to the lungs. It is a *low-pressure circulation* originating from the right ventricle and terminating at the left atrium, which receives oxygenated blood from the lungs.
Double circulation ensures that oxygenated blood does not mix with deoxygenated blood. So, only oxygenated blood is circulated to all the organ systems of the body.

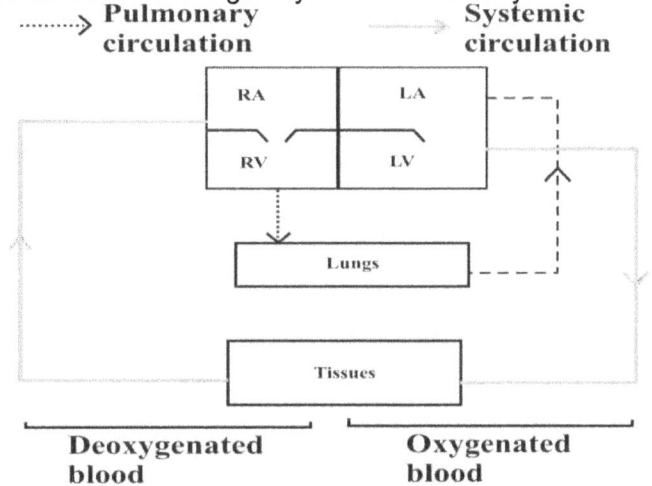

Fig.4.4

The Coronary circulation
The Coronary circulation supplies blood to the cardiac muscles. The coronary artery branches from the aorta and enters the heart muscles, as shown in fig.4.5. It splits into capillaries and exchange of materials takes place between the blood and tissues. The capillaries reunite to form veins, which pour blood back into the right atrium through the coronary sinus. About 5% of the total cardiac output goes to coronary circulation. Blockage of the coronary vessels leads to heart attacks or Coronary heart diseases (CHD).

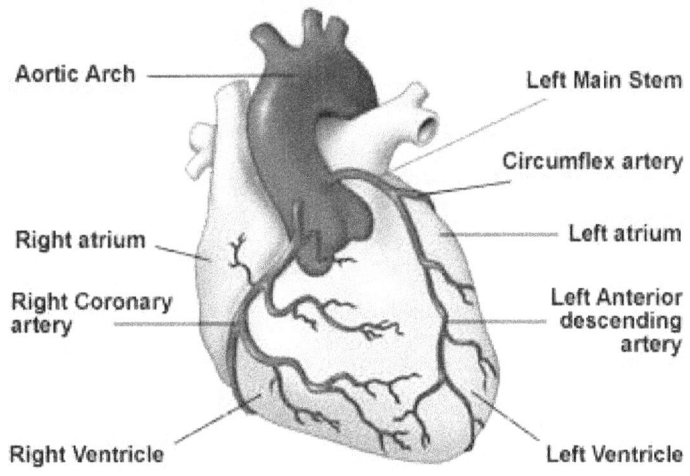

Fig.4.5

The cardiac muscles contract and relax continuously throughout life. Muscle contraction requires large amounts of ATP, which is produced by respiration. So, the cardiac muscle needs a rapid and continuous supply of oxygenated blood to enable ATP production by aerobic respiration. This is made possible by the coronary circulation.

Lack of oxygen in cardiac muscle tissue, due to blocks in the coronary arteries, will result in anaerobic respiration, which produces lactic acid and very small quantities of ATP. The lactic acid will lower the pH inside the muscle cells and deactivates muscle proteins and enzymes, leading to muscle fatigue and cardiac arrest or heart attacks. The lack of ATP will also prevent the muscles from contracting.

The Cardiac Cycle

The cardiac cycle is the rhythmic contraction and relaxation of the atria and ventricles during one complete heartbeat. The cardiac cycle can be divided in to 3 main stages, as shown in fig.4.6.

1) **Atrial systole, ventricular diastole (0.13 sec)** – The atria contract and push blood into the ventricles through atrio-ventricular valves.
2) **Ventricular systole, atrial diastole (0.3 sec)** – The ventricles contract and pumps blood from ventricles into arteries through semi-lunar valves. Atria are being refilled during this time.
3) **Complete diastole (0.4 sec)** – atria are being filled with blood from veins. Blood slowly oozes into ventricles from atria.

Systole means contraction;
Diastole means relaxation.

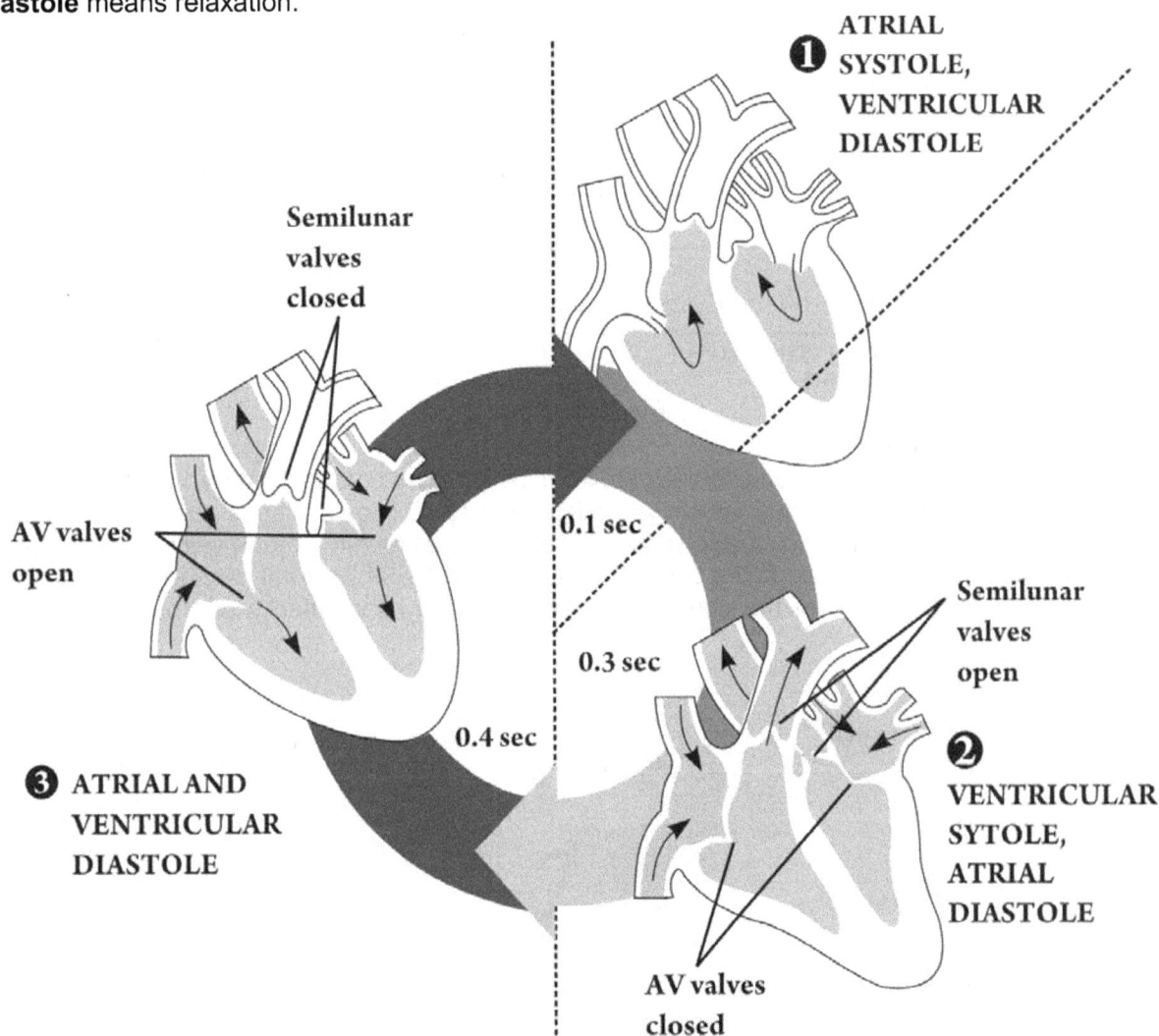

Fig.4.6

OPENING and CLOSING of valves

The semi-lunar valves are open whenever pressure in the ventricles is greater than pressure in the arteries. At all other times semi-lunar valves are closed.

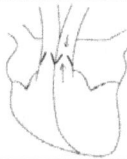

Atria ventricular valves are open whenever the pressure in atria is greater than pressure in ventricles. At all other times the atrio-ventricular valves are closed.

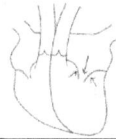

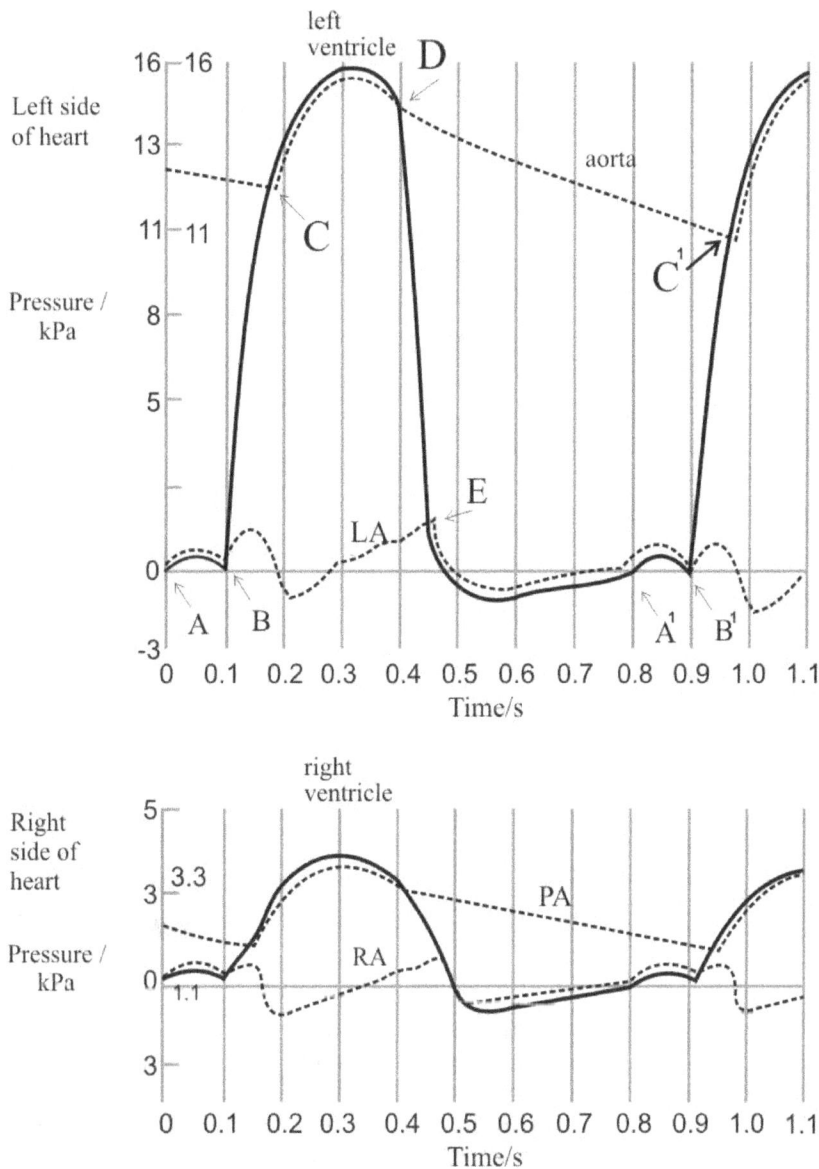

Fig.4.7

The pressure changes in both sides of the heart occur simultaneously. Both ventricles contract together and both atria contract together. However, the pressure in the right side is lower due to the thinner walls of the right ventricle.

Elastic recoil of arteries: During ventricular systole, the aorta is stretched due to the high pressure. During complete diastole, the aortic semilunar valve closes and the aorta walls recoil to maintain a high pressure. This pressure helps to propel the blood forward in the arteries. The elastic recoil of arteries can be felt as the pulse.

The Fig.4.7 shows the pressure changes in the heart. The letters on the graph show various events that occur during one complete cardiac cycle.

A – A¹: one complete heartbeat.

A – B: atrial systole (0 to 0.1s).

B – D: ventricular systole (0.1 to 0.4s).

D – A¹: complete diastole (0.4 to 0.8s).

at B: Ventricular systole begins. Atrio-ventricular valves close. **'Lubb'** sound.

B – C: Ventricular systole continues, but semi-lunar valves remain closed as pressure in aorta is greater than in ventricle.

at C: semi-lunar valves open as pressure in ventricle now exceeds pressure in aorta.

C – D: Ventricular systole continues and blood is pumped into the aorta. Aorta walls are stretched due to the high pressure.

at D: Semi-lunar valves closes with a **'Dubb'** sound. Pressure in ventricle is lower than in aorta as all the blood has been pumped out of ventricle and ventricles are relaxed.

D – C¹: Elastic recoil of aorta walls maintains a high pressure in the aorta and pushes the blood forward into the arteries.

at E: Atrio-ventricular valves open as blood in the atrium pushes the valve open and oozes or trickles into the ventricles.

E – A¹: blood keeps flowing into the ventricles from the atria.

Myogenic stimulation

Myogenic stimulation of the cardiac cycle means that the stimulus for the contraction of cardiac muscles is generated within the heart itself (not generated by brain or spinal cord). This is evident from the fact that isolated cardiac muscle cells will continue to contract and relax rhythmically when placed in an isotonic solution, without any electrical stimulation. This indicates that cardiac muscles are self-excitatory (myogenic). However, the rate of heartbeat can be altered by the nervous system or hormones (like adrenaline).

Coordination of heartbeat (cardiac cycle)

- The impulse for heartbeat originates at the Sino atrial node (SAN).

- This impulse (wave of excitation) spreads through the atrial muscles (intercalated discs of cardiac muscles enhance the transmission of excitation), causing atrial systole.

- The impulse is than taken up by the Atrio-Ventricular Node (AVN) and passed down to the apex of the heart, by the bundle of His.

- The wave of excitation then spreads into the ventricular walls (through the Purkinje fibres) causing ventricular systole.

- The lack of conducting tissue between SAN and AVN causes a time delay. So that ventricular systole begins only after completion of atrial systole.

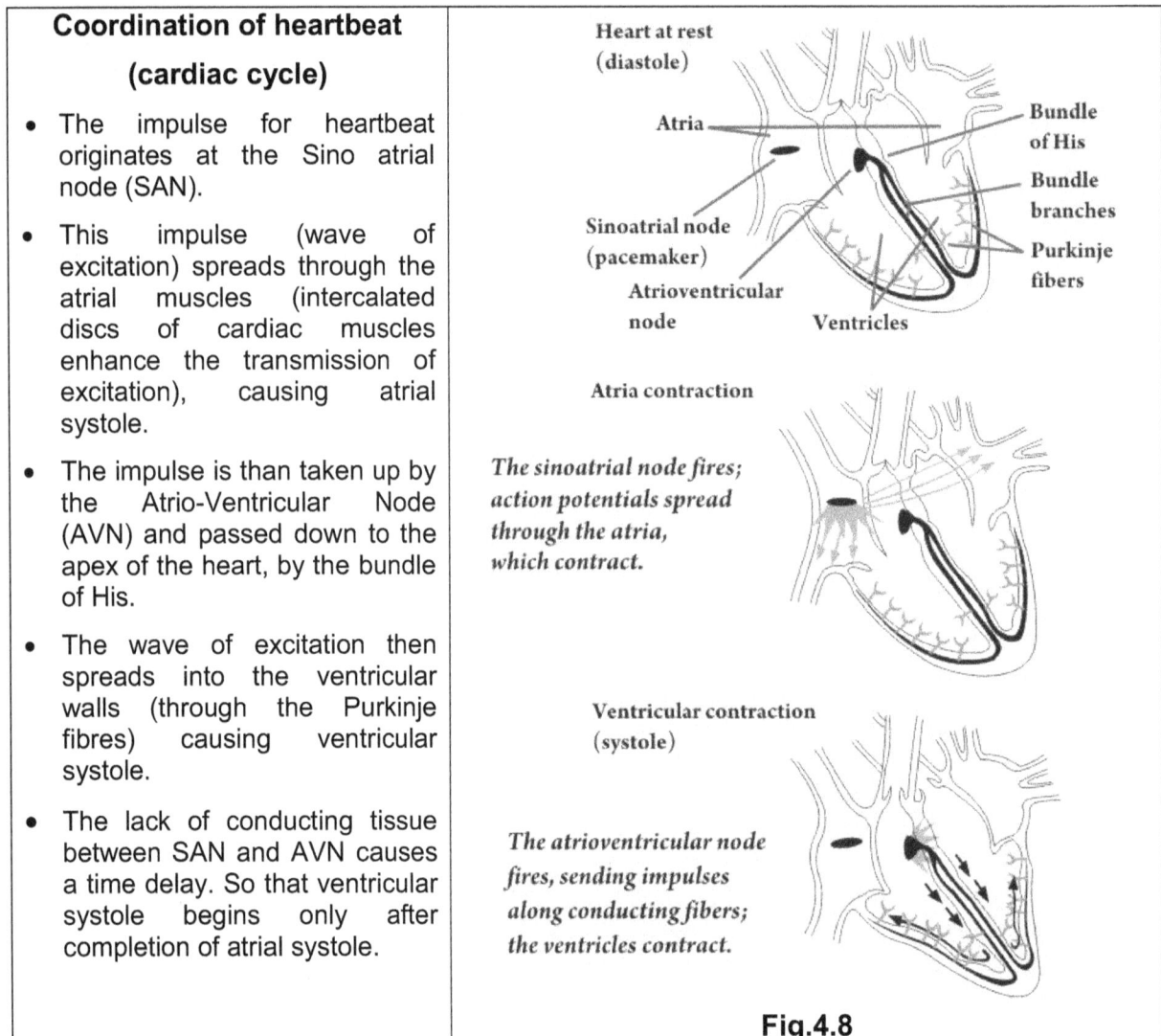

Heart at rest (diastole)

Atria — Bundle of His

Sinoatrial node (pacemaker) — Bundle branches

Atrioventricular node — Ventricles — Purkinje fibers

Atria contraction

The sinoatrial node fires; action potentials spread through the atria, which contract.

Ventricular contraction (systole)

The atrioventricular node fires, sending impulses along conducting fibers; the ventricles contract.

Fig.4.8

Structure of mammalian heart related to its function

1. The walls of the left ventricle are thicker than the right ventricle. This ensures that there is a high pressure in the systemic circulation, as the left ventricle has to pump blood to all parts of the body (even to the head, against the force of gravity). The right ventricle pumps blood to the lungs at a lower pressure, as the lungs are very close to the heart and pulmonary oedema would occur if the pressure is too high.

2. The Sino Atrial Node (SAN) is located on the roof of the atrium, so that the atria contract from the top of the heart and push blood downwards into the ventricles.

3. The Purkinje fibres extend into the ventricle walls from the apex. This ensures that ventricular systole begins from the apex and pushes blood upwards into the arteries. It also results in the closing of the atrio-ventricular valves.

4. The heart muscle has its own blood supply. Even though the heart is filled with blood, the walls of atria and ventricles have too small a surface area to allow diffusion of substances to or from all heart cells. The coronary circulation ensures that every heart muscle cell is close to a capillary for rapid diffusion of oxygen, glucose, nutrients, waste products and carbon dioxide. This helps the cardiac muscle to continue beating rapidly without getting exhausted.

5. There is a non-conducting layer between the atria and ventricles. The impulses can only pass to the ventricles via the AVN, thus causing a delay and ensuring coordinated contraction of the ventricles after the atria have completely contracted.

Blood vessels

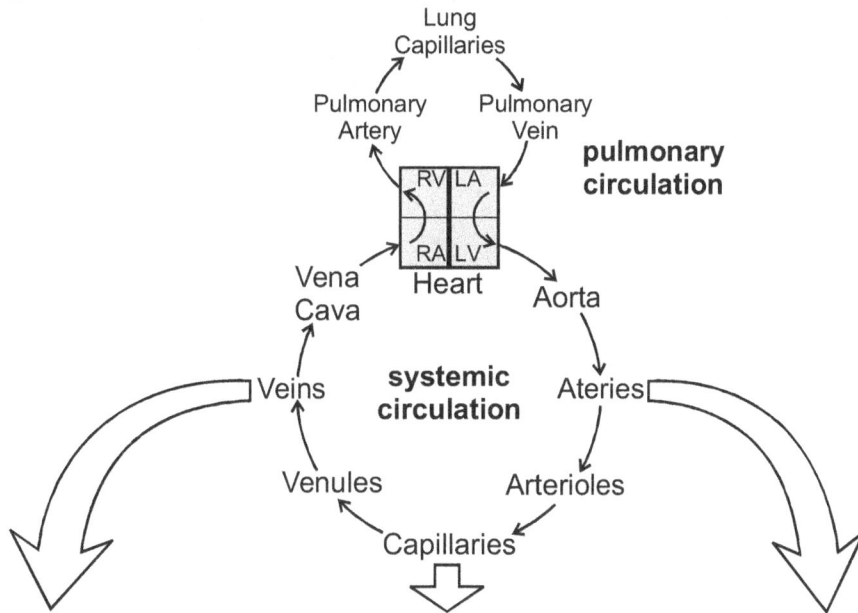

Lung
Capillaries

Pulmonary   Pulmonary
Artery   Vein

pulmonary circulation

RV | LA
RA | LV

Vena   Heart   Aorta
Cava

Veins   **systemic circulation**   Ateries

Venules   Arterioles

Capillaries

Veins and Venules	Capillaries	Arteries and Arterioles
(Tunica Intima) Endothelium — (Tunica Externa) Collagen and connective tissue — (Tunica Media) Smooth muscle and elastic tissue — Semilunar Valve — Lumen (Blood) — 0.1 to 20 mm **Fig.4.9a**	Basement membrane — Endothelium (Tunica Intima) — Red Blood Cell — 8µm **Fig.4.9b**	(Tunica Intima) Endothelium — (Tunica Externa) Collagen and connective tissue — (Tunica Media) Smooth muscle and elastic tissue — Lumen (Blood) — 0.1 to 10 mm **Fig.4.9c**
Function is to carry blood from tissues to the heart	Function is to allow exchange of materials between the blood and the tissues	Function is to carry blood from the heart to the tissues
Thin walls, mainly collagen, since blood at low pressure	Very thin, permeable walls, only one cell thick to allow easy diffusion of materials	Thick walls with smooth elastic layers to resist high pressure and muscle layer to aid pumping(recoil)
Large lumen to reduce resistance to flow.	Very small lumen. Blood cells must distort to pass through.	Small lumen
Many valves to prevent backflow of blood	No valves	No valves (except in heart)
Blood at low pressure due to large lumen	Blood pressure falls in capillaries as volume of blood decreases due to formation of tissue fluid and the large surface area	Blood at high pressure due to narrow lumen and elastic recoil
Blood usually deoxygenated (except in pulmonary vein and umbilical vein)	Blood changes from oxygenated to deoxygenated (except in lungs)	Blood usually oxygenated (except in pulmonary artery and umbilical artery)
Collagen makes vessels strong and durable. **Elastic fibres** allow vessels to stretch and recoil. **Smooth muscle cells** in the walls allow vessels to constrict and dilate.		

Blood flow in veins.

The **skeletal muscles contract to propel the blood** in the veins towards the heart. The **semi-lunar valves prevent the backflow of blood** in the veins, as in fig.4.10. **Breathing movements reduce the pressure in the thoracic cavity** and cause **a suction force** to assist the flow of blood towards the heart.

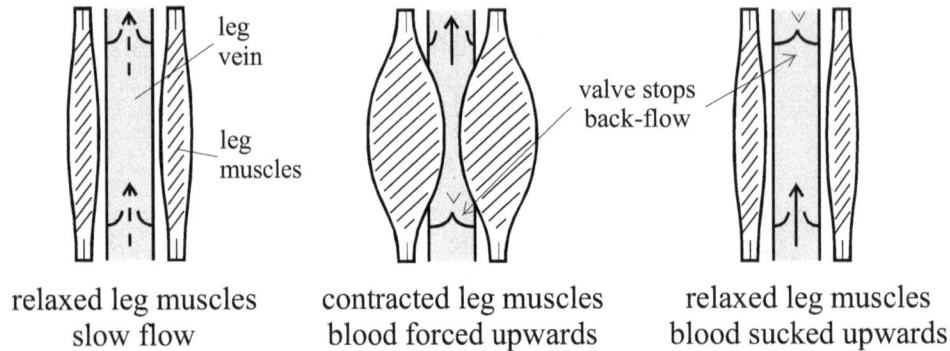

relaxed leg muscles
slow flow

contracted leg muscles
blood forced upwards

valve stops
back-flow

relaxed leg muscles
blood sucked upwards

Fig.4.10

The body relies on constant contraction of skeletal muscles to get the blood back to the heart. This explains why soldiers standing still on parade for long periods can faint, and why sitting still on a long flight can cause swelling of the ankles and Deep Vein Thrombosis (DVT or "economy class syndrome"), where small blood clots collect in the legs.

FORMATION AND REABSORPTION OF TISSUE FLUID

Tissue fluid is a transport medium between blood in the capillaries and tissues. It carries nutrients from capillaries to tissues and other wastes flow back from tissue to capillaries. These substances can pass from capillaries to tissue fluid and vice versa by ultrafiltration, osmosis and diffusion. Tissue fluid is formed from blood at the arterial end of the capillaries. The high blood pressure, thin and porous walls of the capillaries causes the liquid component of blood (except large proteins) to be squeezed out of the capillaries. About 85% of the tissue fluid is reabsorbed at the venous ends of the capillaries mainly by osmosis. Solutes move along the Pressure gradient caused by difference in water potential between the inside and outside of the capillaries. The remaining 15% of tissue fluid is collected by lymph capillaries which pour lymph back into the veins. Oedema is the accumulation of tissue fluid, causing swelling. This may occur due to - Blockage of lymph vessels; Increased permeability of capillaries, Increased in blood pressure; Lack of proteins in the plasma.

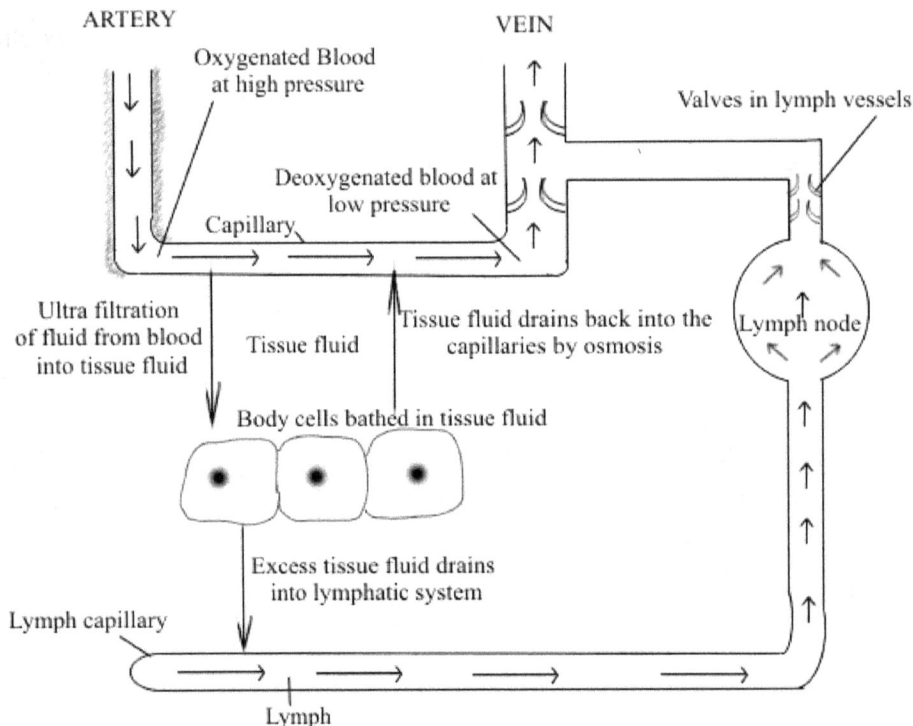

Fig.4.12

Core practical

9 Describe how the effect of caffeine on heart rate in *Daphnia* can be investigated practically, and discuss whether there are ethical issues in the use of invertebrates.

Daphnia, the water flea, is a small freshwater crustacean which lacks physiological methods of maintaining a constant body temperature. This means that if the environmental temperature changes, its body temperature does so too and its metabolic rate will be expected to rise or fall accordingly. So the temperature of the organism must be kept constant during the procedure.

In this investigation we shall test the hypothesis that as the concentration of caffeine changes, the heartbeat rate **(cardiac frequency)** of *Daphnia* also changes. Fortunately Daphnia is relatively transparent and its heart can be seen quite easily under the low Power of the microscope.

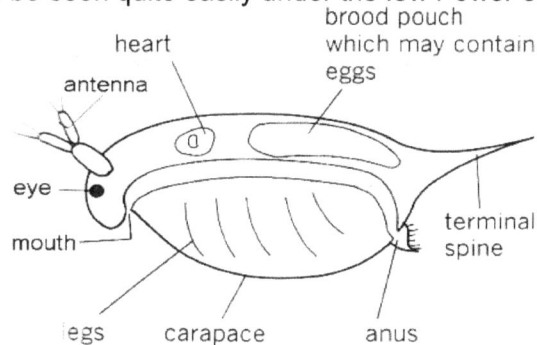

Setting up the experiment

1. Select a large specimen and, with a pipette, transfer it to the centre of a small, dry Petri dish. With filter paper remove excess water from around the specimen so that it is completely stranded.

2. With a seeker place a small blob of silicone grease onto the floor of the Petri dish. Then wipe the needle clean and use it to gently push the posterior end of the animal into the grease so that it is firmly anchored. Now fill the Petri dish with water at 30°C.

3. Place the Petri dish on the stage of a microscope and observe the animal under low Power. The figure above shows the position of the heart, watch it beating. Don't confuse the beating of the heart with the flapping of the legs.

4. Surround the animal with a circular heating coil and fix it in position as shown in the figure below. Also clamp a small mercury thermometer, or the temperature probe of a digital thermometer, into position.

A Petri dish and heating coil viewed from above	B View of set-up from front
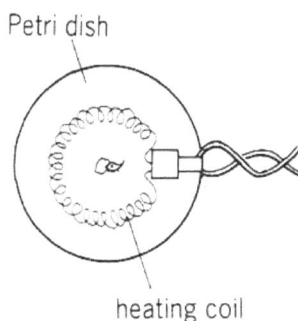	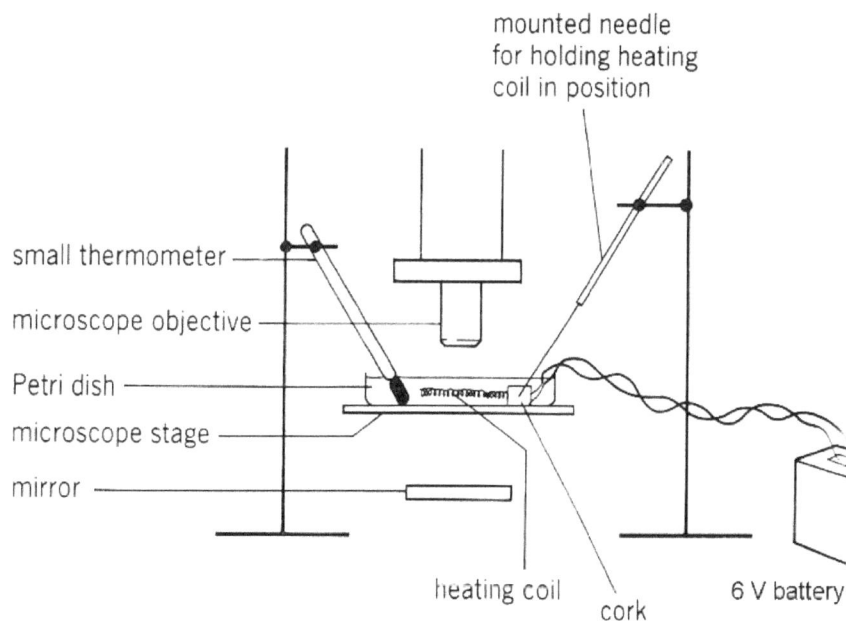

Estimating the cardiac frequency

- A convenient way of doing this is to time how long it takes for the heart to beat 50 times. If it is beating too frequently for every beat to be counted, make a mark on a piece of paper every tenth beat. Do several practice runs to get used to the technique when you feel ready, proceed as follows:

  Replace the distilled water in the Petri dish with caffeine solutions of concentration 1mol dm^{-3} at 30°C. Estimate the cardiac frequency.

- Switch on the heater so that the water gradually warms up. If the temperature of the water rises too rapidly, switch off the heater and, if necessary, add a few ice chippings. Estimate the cardiac frequency at caffeine concentrations of 2mol dm^{-3}, 3mol dm^{-3}, 4mol dm^{-3}, 5 mol dm^{-3} and 6 mol dm^{-3}, noting the temperature each time.

- Present your results in a table and. if you have sufficient readings, draw a graph of the cardiac frequency as a function of the caffeine concentration.

Ethical issues in the use of invertebrates

Animals are often used for research to enhance scientific knowledge. Monkeys are commonly used for brain research, dogs are used in behavioural experiments, rabbits and mice are often dissected in laboratories, mice and fruit flies are used in genetic research, etc. Is it cruel and unfair to utilise these organisms for research? The issue is very controversial and the ethical guidelines vary from country to country and person to person. In the UK it is considered ethical to use invertebrates, such as *Daphnia* in scientific studies, for the following reasons: *Daphnia* has **reduced awareness of pain** because of the lack of a well developed nervous system. It is transparent and its heart is visible **without the need for dissection**. *Daphnia* is **abundant** in nature and there is no threat to it or its dependent species (food chains). Some people also feel that it is bred for fish food and **will thus die anyway**. *Daphnia* can reproduce asexually and may be clones, therefore there is **no loss of genetic variation**.

CHAPTER FIVE
CARDIO-VASCULAR DISEASES

Blood clotting – Thrombosis

Platelets are fragments of cells broken off from large cells (called megakaryocytes) in the bone marrow. They play an important role in blood clotting. The process of blood clotting is illustrated in fig5.1.

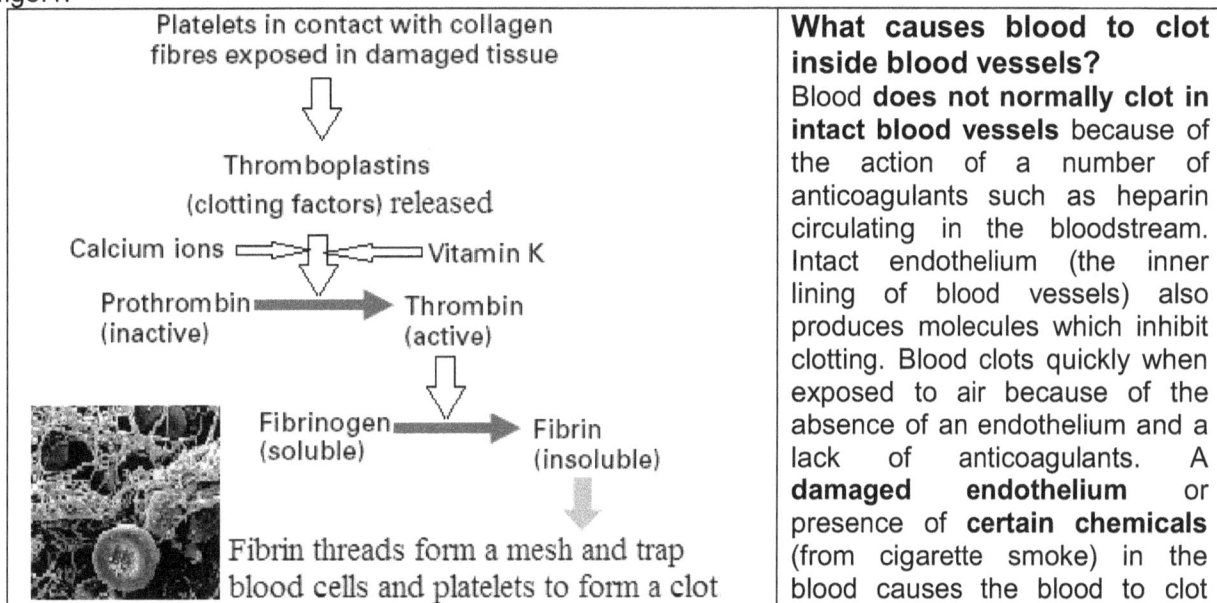

Platelets in contact with collagen fibres exposed in damaged tissue

Thromboplastins (clotting factors) released

Calcium ions ⟹ ⟸ Vitamin K

Prothrombin (inactive) ⟶ Thrombin (active)

Fibrinogen (soluble) ⟶ Fibrin (insoluble)

Fibrin threads form a mesh and trap blood cells and platelets to form a clot

Fig.5.1

Platelets stick to a site of damaged tissue. Thromboplastin is released from this area, which results in a plasma protein called prothrombin being converted to thrombin. Thrombin acts as an enzyme and catalyses the conversion of the soluble plasma protein fibrinogen into long, insoluble strands of the protein fibrin These strands form a mesh that traps red blood cells to form a blood clot. Blood clotting minimizes blood loss following injury. The blood coagulates to form a solid plug (clot) made or cells trapped in a fibrous network. The **clot**

- **prevents further blood loss,**
- **reduces the risk of pathogens (harmful micro-organisms) entering the body, and**
- **provides a framework for the repair of damaged tissue.**

What causes blood to clot inside blood vessels?

Blood **does not normally clot in intact blood vessels** because of the action of a number of anticoagulants such as heparin circulating in the bloodstream. Intact endothelium (the inner lining of blood vessels) also produces molecules which inhibit clotting. Blood clots quickly when exposed to air because of the absence of an endothelium and a lack of anticoagulants. A **damaged endothelium** or presence of **certain chemicals** (from cigarette smoke) in the blood causes the blood to clot (Thrombosis).

Why does blood clot in arteries more often than in veins?

Blood clotting is initiated when the platelets come in contact with collagen in the walls of damaged blood vessels. The platelets get activated and their shape changes. The blood in arteries flows with a high pulsating pressure than in veins. The high pulsating pressure in arteries leads to more damage in the endothelium, increasing the chances atherosclerosis and internal clots.

Blood clotting is an essential process, for the reasons stated above, however if the blood clots inside the arteries then it would increase the risk of suffering from heart attacks and strokes. For example, some people carry the **platelet antigen gene** which causes platelets to be overly active in forming blood clots. These clots could block the arteries and increase the risk of cardiovascular diseases. Other factors, like high blood pressure can damage the endothelium and increase the risk of clotting in arteries, increasing the risk of heart attacks and strokes.

How Atherosclerosis Develops

The wall of an artery is composed of several layers. The lining or inner layer (endothelium) is usually smooth and unbroken. Atherosclerosis begins when the **endothelium is injured or diseased**. Then certain white blood cells called monocytes and T cells are activated and move out of the bloodstream and through the lining of an artery into the artery's wall, this is called as the inflammatory response. Inside the lining, they are transformed into **foam cells**, which are cells that collect fatty materials, mainly cholesterol. In time, smooth muscle cells move from the middle layer into the lining of the artery's wall and multiply there. Connective and elastic tissue materials also accumulate there, as may cell debris, cholesterol crystals, and calcium. This accumulation of fat-laden cells, smooth muscle cells, and other materials forms a patchy deposit called an **atheroma** or **atherosclerotic plaque,** as shown in fig.5.2.

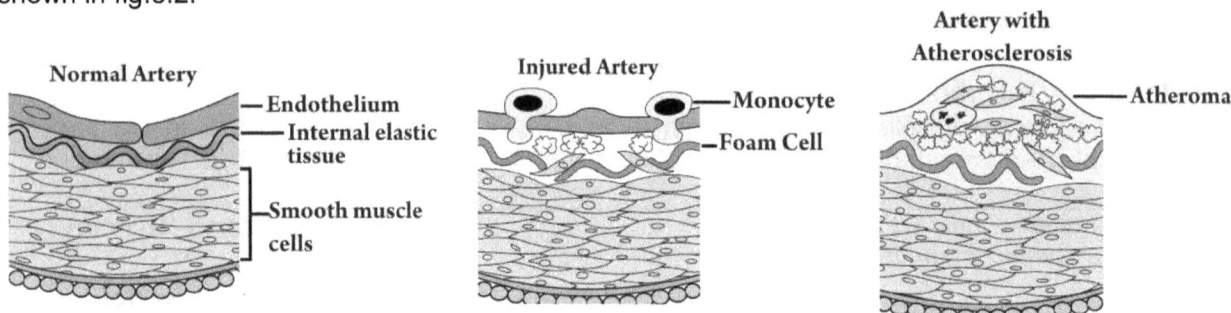

Fig.5.2

As they grow, some plaques thicken the artery's wall and bulge into the channel of the artery. This hardened area means that part of the artery wall **hardens**, so it is **less elastic** than it should be. This is **atherosclerosis.**

The lumen of the artery becomes much smaller and less elastic as a result of the plaque. This increases the blood pressure, making it harder for the heart to pump blood around the body. The raised blood pressure makes damage more likely in the endothelial lining in other areas, and more plaques will form. More plaques will make the blood pressure even higher. This is an example of **positive feedback.**

These plaques may narrow or block an artery, reducing or stopping blood flow. Other plaques do not block the artery very much but may split open, triggering a blood clot that suddenly blocks the artery. Plaques may be scattered throughout medium-sized and large arteries, but they usually form where the **arteries branch** due to **turbulence** of blood flow at these regions.

Atherosclerosis is caused by repeated injury to the walls of arteries. Many factors contribute to this injury, including **high blood pressure, tobacco smoke, diabetes** and **high levels of cholesterol** in the blood.

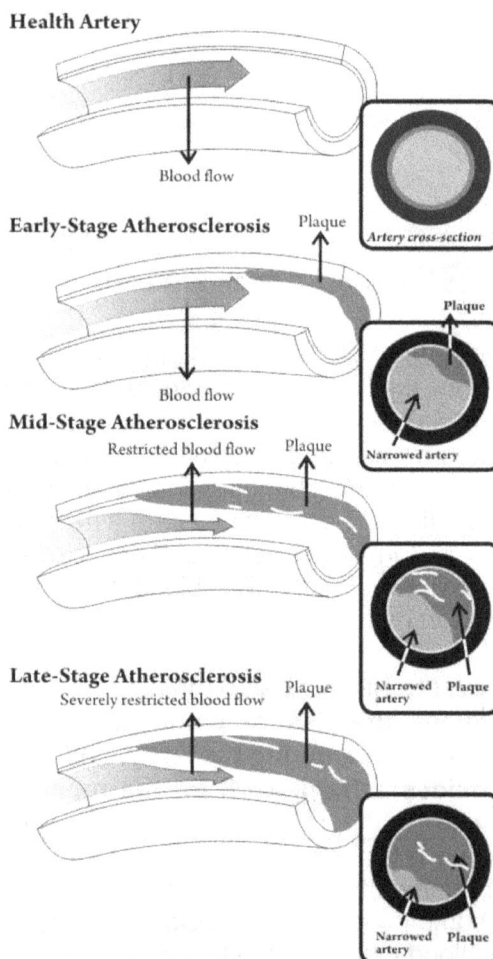

Fig.5.3

Role of blood clotting and atherosclerosis in cardiovascular disease (CVD)

Cardiovascular diseases (CVD) are diseases associated with the coronary circulation or diseases of the major arteries, like arteries in the heart or brain. The sequence of events is shown in fig.5.4.

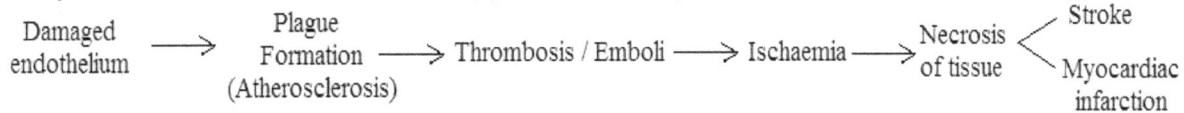

Damaged endothelium $\longrightarrow$ Plague Formation (Atherosclerosis) $\longrightarrow$ Thrombosis / Emboli $\longrightarrow$ Ischaemia $\longrightarrow$ Necrosis of tissue $<$ Stroke / Myocardiac infarction

Fig.5.4

Coronary heart disease (CHD)

If blood clotting and atherosclerosis occurs inside a coronary blood vessel, then the vessel gets blocked and results in lack of blood flow to the cardiac muscles. The clotting of blood in the coronary artery is called as **coronary thrombosis**. This will lead to lack of oxygen and glucose in the cardiac muscle tissue. This is referred to as **ischaemia**. Ischaemia means local anaemia in a given body part sometimes resulting from vasoconstriction, **thrombosis** or **embolism**. Anaemia is a deficiency of red blood cells. The anaerobic conditions will cause the muscles to stop working and damaged cells die. Lactic acid builds up in the cardiac muscles and results in pain called as **angina pectoris**. This may also lead to myocardial infarction. **Myocardial infarction is** a heart attack. Abbreviated MI. The word "infarction" comes from the Latin "infarcire" meaning "to plug up or cram." It refers to the clogging of the artery. The term "myocardial infarction" focuses on the myocardium (the heart muscle) and the changes that occur in it due to the sudden deprivation of circulating blood. The main change is necrosis (death) of myocardial tissue.

Cerebrovascular diseases (Lead to strokes)

If blood clotting occurs inside an artery supplying blood to the brain, then it can result in stroke. The stroke caused by the clotting of blood inside the artery is called as an ischaemic stroke.

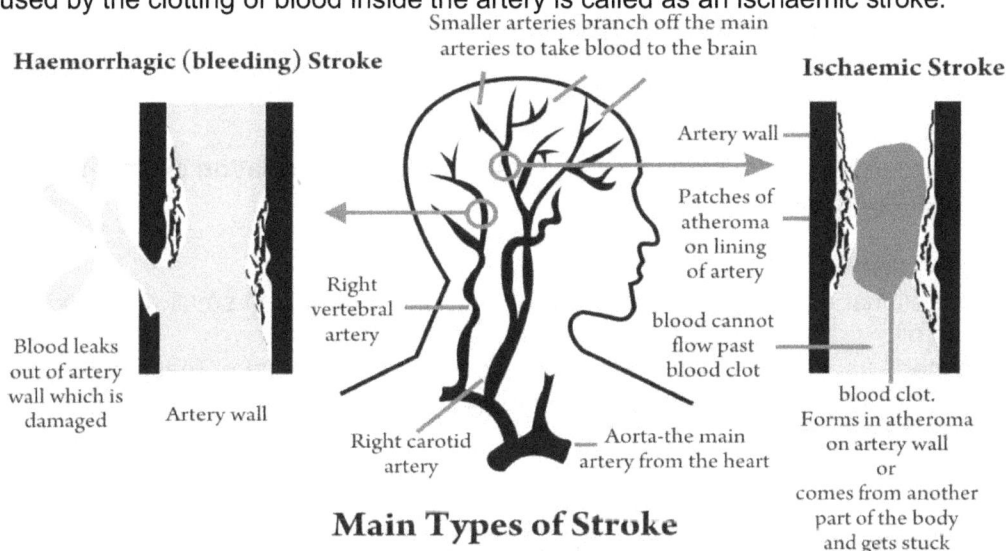

Haemorrhagic (bleeding) Stroke

Smaller arteries branch off the main arteries to take blood to the brain

Ischaemic Stroke

Artery wall

Patches of atheroma on lining of artery

blood cannot flow past blood clot

Right vertebral artery

Blood leaks out of artery wall which is damaged

Artery wall

Right carotid artery

Aorta-the main artery from the heart

blood clot. Forms in atheroma on artery wall or comes from another part of the body and gets stuck

Main Types of Stroke

Fig.5.5

The symptoms of strokes vary, depending on how much of the brain is affected. Very often the blood is cut off from one part or one side of the brain only. Symptoms include dizziness, confusion, slurred speech, blurred vision or loss of part of the vision (usually just in one eye) and numbness. In more severe strokes there can be paralysis, usually just down one side of the body.

Peripheral arterial disease (PAD): This occurs when plaque builds up in the major arteries that supply oxygen-rich blood to the legs, arms, and pelvis. When blood flow to these parts of your body is reduced or blocked, it can lead to numbness, pain, and sometimes dangerous infections.

Aneurysm

An aneurysm is a bulge in the artery wall. This occurs when blood tends to build up behind a blockage in the artery. This causes the artery wall to be severely weakened. The weakened artery may split open, leading to severe internal haemorrhage (bleeding). If an aneurysm forms in the brain, it could lead to stroke. If it occurs in the dorsal aorta (in the abdomen) it could lead to a drop in blood pressure, which could be fatal.

CHAPTER SIX
RISK FACTORS

Learning outcomes: by the end of this chapter you should be able to
Edexcel Syllabus Spec 12: *Describe the factors that increase the risk of CVD (genetic, diet, age, gender, high blood pressure, smoking and inactivity).*

Measuring or estimating risk

A risk factor is anything that affects or raises the **chance** of harm. **Chance** or **probability** can be measured in terms of percentage or ratio.

Risk describes the probability that a particular event will happen. Probability means the chance or likelihood of the event, calculated mathematically.

In 2005, 335969 people died due to Coronary Heart Diseases (CHD). The total UK population at that time was 60 209 408, so we can calculate the average risk in a year of someone in the UK dying from Coronary Heart Diseases (CHD) as:

$$Risk = (Number\ of\ events\ /\ total\ entities)\ x\ 100$$
$$Risk = (335969\ /\ 60\ 209\ 408)\ x\ 100 = 0.55\%$$

Or

Risk may also be expressed as a ratio or probability. For example the risk of people dying from CHD in 2005, in the UK population is 1 in 179, i.e. (335969 in 60 209408).

Risk factors

Any factor which increases the chances of suffering from cardiovascular diseases is called as a risk factor.

Diet

The following recommendations are made by the **British Nutrition Foundation** in order to reduce the risk of cardiovascular diseases.

Eat a healthy, varied diet
- Use vegetable oil that is high in unsaturated fat in cooking, but only in small amounts, *e.g.* olive oil or rapeseed oil.
- Choose lean meat, poultry, beans and alternatives instead of fatty meat or meat products.
- Choose low-fat dairy foods, like skimmed and semi-skimmed milk or low-fat yoghurt.
- Eat less fat and fatty foods

Explanation

Polyunsaturated fatty acids (Vegetable oils)	Saturated fatty acids (Animal fats)
Vegetable oil contains high percentage of polyunsaturated fatty acids, which form HDLs.	Animal fat contains high percentage of saturated fatty acids, which form LDLs.
Hence, vegetable oils (HDLs) reduce blood cholesterol and reduce the chance of forming atherosclerosis.	Hence, animal fats (LDLs) contribute to the build up of atheroma or plaque, causing atherosclerosis.

Lipid rich food also contains a lot of energy. All this energy may not be used by the body and the excess will be stored in adipose tissue. This increases the risk from **obesity related** disorders.

Margarine is a low fat spread enriched with plant sterols. Plant sterols have been clinically proven to lower cholesterol levels by reducing the cholesterol being absorbed from the gut. This helps to maintain a healthy heart. Independent clinical trials prove that moving to a diet that includes 20-25g of healthier margarine per day gives average LDL cholesterol reductions of 10—15%. There is no effect on HDL cholesterol.

- Eat more fruits and vegetables, at least 5 portions a day.
- Eat more starchy foods, like potatoes, rice, pasta, bread and breakfast cereals.
- Choose high fibre, whole-grain products.

Explanation

Fruits and vegetables contain a lot of fibre and antioxidants like Vitamin C and beta carotene. The antioxidants neutralise free radicals and prevent cellular damage. Free radicals are charged particles produced by metabolism in the body. They attach to cells and cause cellular damage. Antioxidants can neutralise these radicals and reduce the harm caused to the cells (specially myocardial cells and endothelium).

- Eat fish at least twice a week, of which one portion should be oily fish.

Explanation

Fish oil contains *omega-3 fatty acids*, a group of polyunsaturated fatty acids with their first bond between the third and fourth carbon atoms. These fatty acids are essential for cell functioning and have been linked to a reduction in heart disease and joint inflammation.

- Choose low-salt products and use less salt in cooking. Content should not be more than 6g per day for an adult.

Explanation

Excess salt can cause an increase in the retention of water in the blood by the kidneys. This can increase the volume of blood and blood pressure, increasing the risk of Cardiovascular diseases.

- If you drink alcohol, do so moderately. The guidelines for safe intakes are no more than 3-4 units per day for men and 2-3 units per day for women (1 unit = 8g).

Explanation

Heavy drinking raises blood pressure, contributes to obesity and can cause irregular heartbeat. It also damages liver cells.

Moderate consumption of alcohol has a beneficial effect on health. It is thought to increase the HDL levels in blood. Red wine also contains antioxidants, which reduce the risk of cardiovascular diseases.

Genetic factors

Familial hypercholesterolemia – increasing risk in children and teenagers

In some cases, high cholesterol runs in families. This is called **familial hypercholesterolemia (FH)**. FH is characterized by high cholesterol levels, specifically very high low-density lipoprotein (LDL, "bad cholesterol") levels, in the blood and early cardiovascular disease (CVD).

FH is caused due to a defective *LDLR* gene that encodes the LDL **Receptor** protein, which normally removes LDL from the circulation, as shown in fig.6.1.

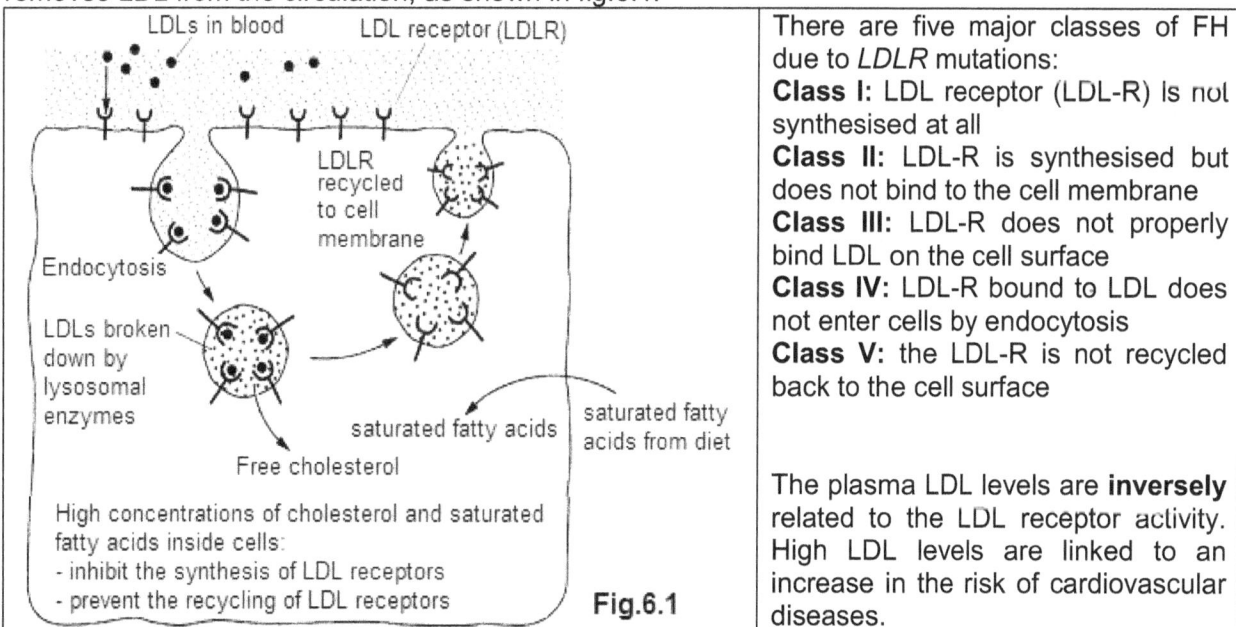

There are five major classes of FH due to *LDLR* mutations:

Class I: LDL receptor (LDL-R) is not synthesised at all

Class II: LDL-R is synthesised but does not bind to the cell membrane

Class III: LDL-R does not properly bind LDL on the cell surface

Class IV: LDL-R bound to LDL does not enter cells by endocytosis

Class V: the LDL-R is not recycled back to the cell surface

The plasma LDL levels are **inversely** related to the LDL receptor activity. High LDL levels are linked to an increase in the risk of cardiovascular diseases.

Fig.6.1

APOE(4) and APOE(2) Genes

Another cluster of genes that is associated with coronary heart disease is the apolipoprotein gene. Apolipoproteins are the protein component of lipoproteins. They are formed in the liver and intestine. The main role is in stabilizing the structure of lipoproteins, shown in fig.6.4 , and recognizing receptors involved in lipoprotein uptake on the plasma membrane of most cells in the body.

One form of the APOE gene (E2) produces a protein that helps lower cholesterol in the blood reducing the risk of developing CVD. Another form (E4) is less effective at removing cholesterol so increases the risk of CVD.

The platelet antigen gene linked to sudden death in athletes.

The platelet antigen gene is a form of a gene which greatly increases the chances of a heart attack. This form of the gene causes platelets to be overly active in forming blood clots and may cause cholesterol to bind to endothelial cells lining blood vessels. Increased deposition of cholesterol in the walls of coronary arteries will lead to more rapid development of atheroma and narrowing of the blood vessel. As the blood slows, the sticky platelets are more likely to form a clot. Any damage to the vessel walls will result in rapid formation of a clot, leading to sudden death.

Age and Gender

Age is a major risk factor for cardiovascular disease including heart disease. The incidence of heart disease increases with age for both men and women. Men develop heart disease, particularly coronary artery disease, at younger ages than do women. Before age 60, one out of three men has heart disease, but only one out of ten women does. After menopause, the incidence of heart disease in women also rises and eventually virtually mirrors heart disease rates in men. Though heart disease once was regarded as a "man's disease," it is the leading cause of death in both men and women.

The following changes take place as the person ages:

Narrowing of arteries and reduced elasticity cause thickening of the left ventricle: Researchers have noted that the wall of the left ventricle of the heart becomes thicker with age. This thickening allows the heart to pump stronger. As our blood vessels age, they become narrower, causing blood pressure to increase. The heart compensates for this by becoming stronger and pumping with more force.

Exercise Capacity Shrinks: As the heart ages, it becomes less able to respond rapidly to chemical messages from the brain. Researchers do not know exactly why the heart does not respond as fast to messages to speed up and adjust to increased activity. The result is the body cannot exercise as long or as intensely as before. This shows up as shortness of breath, a sign that oxygen-rich blood is not moving fast enough through the body because the lungs are trying to breath in more oxygen.

Before menopause, a woman's own oestrogen helps protect her from heart disease by increasing HDL (good) cholesterol and decreasing LDL (bad) cholesterol. After menopause, women have higher concentrations of total cholesterol than men do. But this alone doesn't explain the sudden rise in heart disease risk after menopause. Elevated triglycerides are an especially powerful contributor to cardiovascular risk in women. Low HDL and high triglycerides appear to be the major factor that increase the risk of death from heart disease in women over age 65.

Inactivity - Lack of exercise

Lack of exercise is accepted as a major risk of CHD. As little as, 8 to 16 km of walking or jogging each week can offer some protection against the disease. Aerobic exercise strengthens the heart, enabling it to contract more forcefully so that, for any given workload, the stroke volume increases and the heart rate decreases. Exercise does not guarantee protection from CHD - it just reduces the risk. Exercise also increases HDL levels. Physical activity burns fat, reducing the risk of obesity and plaque formation in blood vessels.

Cigarette smoking

The effects of smoking are shown in fig.6.2.

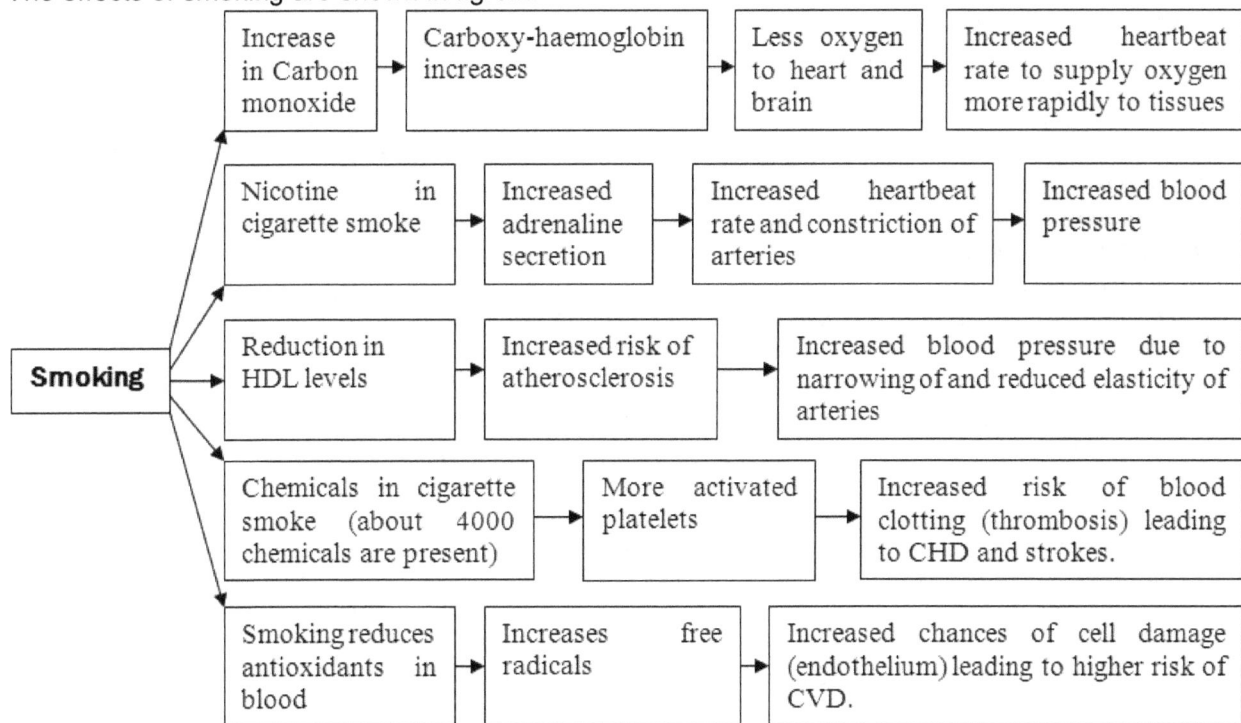

Increase in Carbon monoxide →	Carboxy-haemoglobin increases →	Less oxygen to heart and brain →	Increased heartbeat rate to supply oxygen more rapidly to tissues
Nicotine in cigarette smoke →	Increased adrenaline secretion →	Increased heartbeat rate and constriction of arteries →	Increased blood pressure
Reduction in HDL levels →	Increased risk of atherosclerosis →	Increased blood pressure due to narrowing of and reduced elasticity of arteries	
Chemicals in cigarette smoke (about 4000 chemicals are present) →	More activated platelets →	Increased risk of blood clotting (thrombosis) leading to CHD and strokes.	
Smoking reduces antioxidants in blood →	Increases free radicals →	Increased chances of cell damage (endothelium) leading to higher risk of CVD.	

Smoking (branches to all the above rows)

Fig.6.2

High blood pressure

Blood pressure is usually expressed in terms of the systolic pressure over diastolic pressure, for example 120/80. If the pressure remains consistently higher than normal for an extended period of time, then it is referred to as hypertension or high blood pressure. **High blood pressure is the No. 1 modifiable risk factor for <u>stroke</u>**. It also contributes to heart attacks, heart failure, kidney failure and atherosclerosis. In some cases, it can cause blindness. The relationship of blood pressure levels to the risk of cardiovascular disease is continuous, consistent and independent of other risk factors. That means the higher your blood pressure, the greater your risk of heart attack, heart failure, stroke and kidney disease. The factors affecting blood pressure are shown in the illustration below:

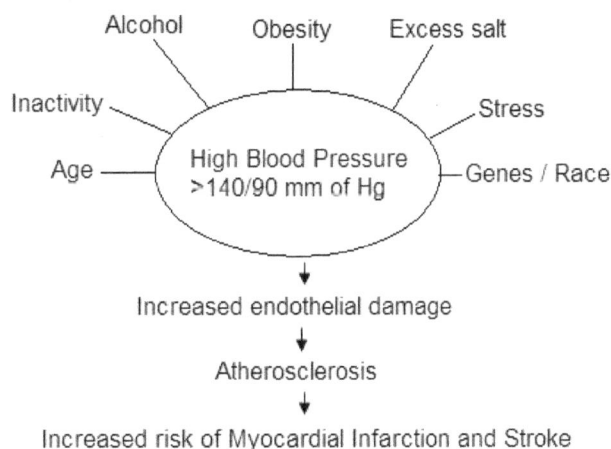

Alcohol   Obesity   Excess salt

Inactivity   Stress

Age — **High Blood Pressure >140/90 mm of Hg** — Genes / Race

↓

Increased endothelial damage

↓

Atherosclerosis

↓

Increased risk of Myocardial Infarction and Stroke

Fig.6.3

Multifactorial disease

A multifactorial disease is one whose development is affected by a range of factors, for example heredity, physical environment, social environment and lifestyle choices all contribute to the risk of developing CVD. Each factor increases the risk of suffering from CVD, but caution must be taken in making conclusions about the cause of CVD. For example, lack of physical activity simply increases your risk of CVD, but **<u>does not CAUSE</u>** CVD.

The cholesterol story (not needed for exams but helps to understand the content better.)

*Cholesterol is a fat-like substance called a lipid that is found in all body cells. Cholesterol in the body comes from two sources. A small amount s eaten in certain foods but most is actually produced by our Liver, and we need some in our bodies to stay healthy. The liver produces cholesterol when **saturated fats** are eaten. Today's diet contains lot of saturated fats and therefore the liver produces more cholesterol than the body actually requires. Cholesterol travels to cells through the bloodstream in special carriers called **lipoproteins**. Two of the most important lipoproteins are **low-density lipoprotein** (LDL – carries bad cholesterol) and **high-density lipoprotein** (HDL – carries good cholesterol). Although we often blame the cholesterol found in foods that we eat for raising blood cholesterol, the main culprit is the **saturated fat** in our food. Foods rich in saturated fat include butter fat in milk products, fat from red meat, and tropical oils such as coconut oil.*

Low-density Lipoprotein

*LDL particles deliver cholesterol to the body's cells. LDL cholesterol is often called "bad cholesterol" because high levels are thought to lead to heart disease and diseases of the blood vessels. Too much LDL in the blood causes fatty plaque to form on artery walls, which starts the process of **atherosclerosis**. The LDLs circulate in the blood stream and bind to receptors on the cell surface membrane before being taken up by the cells. Excess LDL in the diet will overload the membrane receptors and result in high blood cholesterol levels. This may increase the risk of atherosclerosis. Sometimes, an underactive thyroid (called hypothyroidism) may also increase LDL levels.*

High-density Lipoprotein

HDL particles carry cholesterol from the body's cells back to the liver, where it can be removed from the body. HDL is known as "good cholesterol" because high levels are thought to lower the risk of heart disease. Low HDL is often the result of physical inactivity, obesity, or smoking. HDLs carry unsaturated fatty acids and help to lower blood cholesterol levels. It also helps to remove fatty plaques of atherosclerosis.

Triglycerides

Triglycerides are fats that provide energy for the muscles. Like cholesterol, they are delivered to the body's cells by lipoproteins in the blood. Eating foods with a lot of saturated fat or carbohydrates will raise triglyceride levels. Elevated levels are thought to lead to a greater risk of heart disease, but scientists do not agree that high triglycerides alone are a risk factor for heart disease.

The most important factor in raising blood cholesterol levels is eating foods high in saturated fat.

Background information

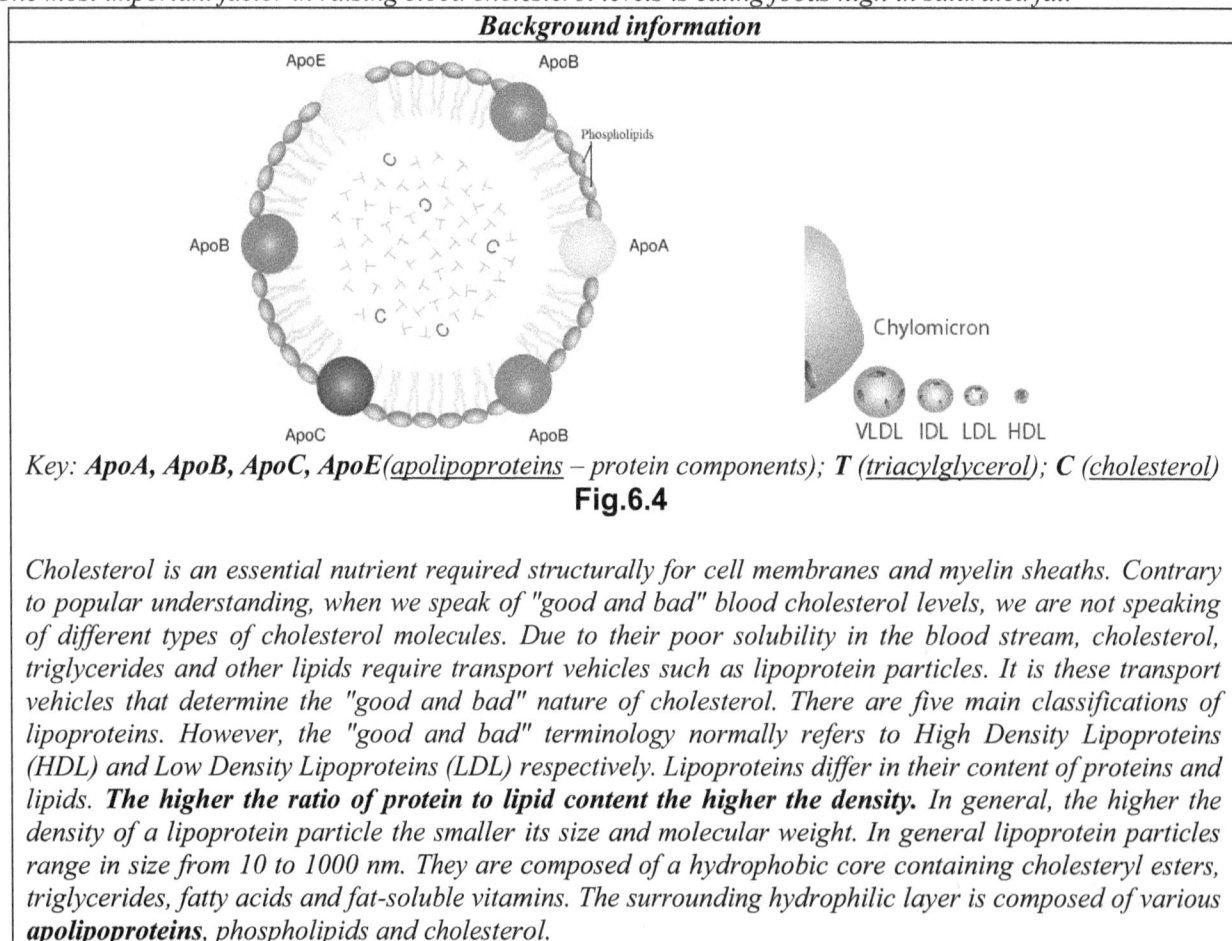

*Key: **ApoA, ApoB, ApoC, ApoE**(apolipoproteins – protein components); **T** (triacylglycerol); **C** (cholesterol)*

Fig.6.4

*Cholesterol is an essential nutrient required structurally for cell membranes and myelin sheaths. Contrary to popular understanding, when we speak of "good and bad" blood cholesterol levels, we are not speaking of different types of cholesterol molecules. Due to their poor solubility in the blood stream, cholesterol, triglycerides and other lipids require transport vehicles such as lipoprotein particles. It is these transport vehicles that determine the "good and bad" nature of cholesterol. There are five main classifications of lipoproteins. However, the "good and bad" terminology normally refers to High Density Lipoproteins (HDL) and Low Density Lipoproteins (LDL) respectively. Lipoproteins differ in their content of proteins and lipids. **The higher the ratio of protein to lipid content the higher the density.** In general, the higher the density of a lipoprotein particle the smaller its size and molecular weight. In general lipoprotein particles range in size from 10 to 1000 nm. They are composed of a hydrophobic core containing cholesteryl esters, triglycerides, fatty acids and fat-soluble vitamins. The surrounding hydrophilic layer is composed of various **apolipoproteins**, phospholipids and cholesterol.*

Learning outcomes: by the end of this chapter you should be able to
Edexcel Syllabus Spec 13: Describe the benefits and risks of treatments for CVD (antihypertensives, plant statins, anticoagulants and platelet inhibitory drugs).

Treatment of atherosclerosis
Treatments for atherosclerosis may include lifestyle changes, medicines, and medical procedures or surgery.

Drugs for treatment of CVD
Many medicines are used to treat patients with cardiac disease. However, individual patients may respond differently to these medications. The effectiveness of a medication may also change with time. For this reason, doctors often prescribe different medications to control, hypertension, for example, until they find a medicine or combination of medicines that works for a given patient. Commonly used medications include: antihypertensives, statins, anticoagulants and platelet inhibitory drugs.

Antihypertensives
Antihypertensives are used to treat high blood pressure (hypertension). The most common types include ARB, ACE inhibitors, diuretics, beta blockers and calcium channel blockers.

Fig.7.3 shows the mechanism by which changes in blood pressure is regulated.

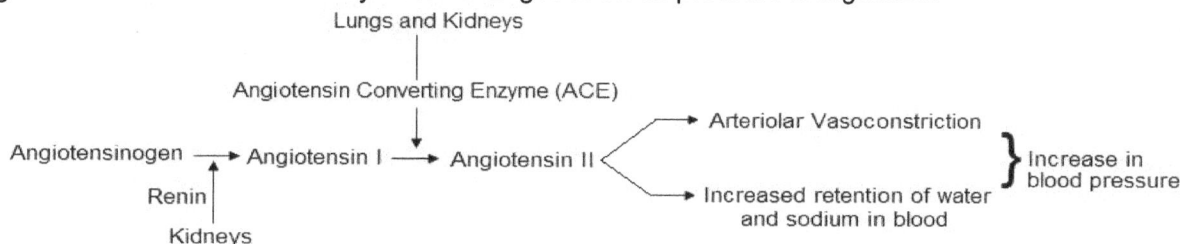

Fig.7.1

The table below summarises the effects of antihypertensives and the risks and benefits.

Mode of action	Benefits	Risks
a. Angiotensin Receptor Blockers (A2RBs) Angiotensin is a substance produced by the liver and kidneys collectively. It causes vessels to constrict by binding with receptors on smooth muscles of the arteries. This elevates the blood pressure. Angiotensin receptor blockers (ARBs) block the action of angiotensin. So, ARBs lower the blood pressure by preventing constriction of the blood vessels.	ARBs are used for controlling high blood pressure, treating heart failure, and preventing kidney failure in people with diabetes or high blood pressure.	Dizziness, headache, drowsiness, diarrhoea, abnormal taste sensation (metallic or salty taste), and rash.
b. ACE Inhibitors ACE (angiotensin converting enzyme - produced by the lungs and kidneys) is used in the formation of angiotensin. **ACE(angiotensin converting enzyme) inhibitors** are used to treat high blood pressure and weakened heart muscles. ACE inhibitors slow (inhibit) the activity of the enzyme (angiotensin converting enzyme), which produces angiotensin. As a result, the	ARBs are used for controlling high blood pressure, treating heart failure, and preventing kidney failure in people with diabetes or high blood pressure.	Skin rash, hay-fever-like symptoms (sneezing, blocked or runny nose, itchy eyes), swelling of your sinuses (sinusitis), sore throat or dry cough, feeling sick or

		vomiting, indigestion
c. Diuretics Diuretics act on the kidneys to help rid the body of excess water and sodium and consequently lower the blood pressure as well.	Diuretics are commonly used to treat high blood pressure (hypertension), congestive heart failure (CHF), and water retention and swelling.	Weakness, Muscle cramps, Skin rash, Diarrhoea, muscle cramps and dizziness.
d. Beta-Blocking Drugs Beta-adrenergic receptors are receptors found on many cells in the body. These receptors bind with the hormone adrenaline. This allows adrenaline to cause constriction of the arteries and to increase the heartbeat rate. Beta-blockers block the beta-adrenergic receptors (beta-adrenergic blocking drugs) and slow the heart rate and decrease blood pressure. They do this by blocking the effects of adrenaline on the body's beta receptors. This slows the nerve impulses that travel through the heart, thus easing the heart's pumping action and widening blood vessels.	Beta-blockers are used to treat a variety of conditions including high blood pressure (hypertension), chest pain (angina), arrhythmias, congestive heart failure (CHF), and mitral valve prolapse. These drugs are specially useful to treat angina caused by stress or psychological shock.	Fatigue, depression, memory loss, dizziness, diarrhoea or constipation.
e. Calcium Channel Blocking Agents Also called calcium antagonists or calcium blockers, these medications affect the movement of calcium into the cells of the heart and blood vessels. As a result, they relax blood vessels and increase the supply of blood and oxygen to the heart, while reducing its workload. Calcium-channel blockers can cause a big drop in blood pressure. This can cause dizziness while standing up. So, after taking a dose, patients need to get up slowly and stay next to your chair or bed until the dizziness wears off.	They are prescribed to treat chest pain (angina), high blood pressure (hypertension) and some irregular heartbeats (arrhythmias). They are not usually prescribed for people with heart failure or other structural damage to the heart.	Calcium-channel blockers can also slow the heart down and cause headaches, constipation, flushing and fluid retention.

Cholesterol lowering drugs – plant statins

Cholesterol is a lipid. It is an important component of cell membranes and it is the starting point for the synthesis of a number of sex hormones. About half of our daily cholesterol requirement is absorbed from the gut. The remainder is synthesised in the liver. The fig.7.4 shows the formation of cholesterol in the liver.

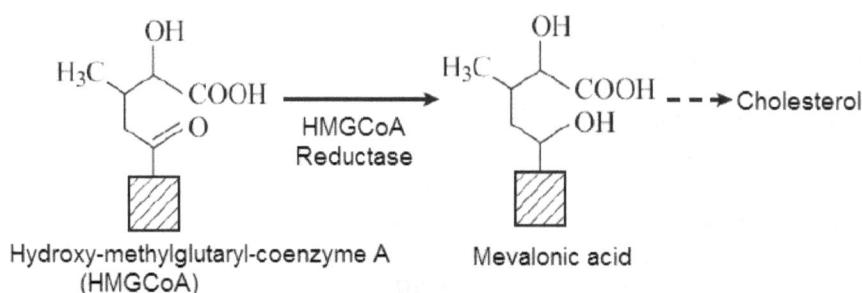

Hydroxy-methylglutaryl-coenzyme A (HMGCoA) → HMGCoA Reductase → Mevalonic acid --→ Cholesterol

Fig.7.2

High blood cholesterol concentration is linked to an increased risk of heart attacks and strokes. Blood cholesterol concentration may be reduced by a suitable diet or statins which lower cholesterol concentration.

Statins block the enzyme in the liver that is responsible for making cholesterol. This enzyme is called hydroxy-methylglutaryl-coenzyme A reductase (HMG-CoA reductase for short). Scientifically, statins are referred to as HMG-CoA reductase inhibitors.

Benefits of statins
Statins are used to lower the blood cholesterol (LDL) levels and reduce the risk of CVD. Doctors often prescribe statins if diet, weight loss, and exercise don't work to sufficiently lower a person's cholesterol (LDL).

Risks of using statins
- Muscle and joint aches (most common)
- Nausea
- Diarrhea
- Constipation
- **Liver damage (rarely)**

Anticoagulants and platelet inhibitory drugs
Anticoagulants are drugs used to prevent clot formation or to prevent a clot that has formed from enlarging. They inhibit clot formation by blocking the action of clotting factors or platelets.

Anticoagulant drugs fall into three categories:
- **Inhibitors of clotting factor synthesis**. These anticoagulants inhibit the production of certain clotting factors in the liver. One example is warfarin, which interferes with the manufacture of prothrombin in the body. Low prothrombin levels make the blood clot less easily.
- **Inhibitors of thrombin.** Thrombin inhibitors interfere with blood clotting by blocking the activity of thrombin. One example is heparin, which inhibits the formation of clotting factors.
- **Antiplatelet drugs.** Antiplatelet drugs interact with platelets, which are fragments of cells that trigger bloodclotting. These drugs block platelets from aggregating into harmful clots. One example is **aspirin**.

Benefits of anticoagulants and platelet inhibitory drugs
Anticoagulant drugs reduce the ability of the blood to form clots. Although blood clotting is essential to prevent serious bleeding in the case of skin cuts, clots inside the blood vessels block the flow of blood to major organs and cause heart attacks and strokes. Although these drugs are sometimes called blood thinners, they do not actually thin the blood. Furthermore, this type of medication will not dissolve clots that already have formed, although the drug stops an existing clot from worsening. However, another type of drug (urokinase), used in **thrombolytic therapy**, will dissolve existing clots.

Anticoagulant drugs are used for a number of conditions. For example, they may be given to prevent blood clots from forming after the replacement of a heart valve or to reduce the risk of a **stroke** or another **heart attack** after a first heart attack. They are also used to reduce the chance of blood clots forming during open heart surgery or bypass surgery. Low doses of these drugs may be given to prevent blood clots in patients who must stay in bed for a long time after certain types of surgery.

Risks of using anticoagulants and platelet inhibitory drugs
Because anticoagulants affect the blood's ability to clot, they can increase the risk of severe bleeding and heavy blood loss. It is thus essential to take these drugs exactly as directed and to see a physician regularly as long as they are prescribed.

The risks of taking aspirin are well known - it irritates the stomach lining and causes bleeding in the stomach which can become serious. A combination of aspirin and clopidogrel can reduce the risk of developing a range of cardiovascular diseases by 20—25% in some low-risk patients. The balance of preventing the blood from clotting too easily while allowing it to clot when necessary is a very fine one. For example, when people are treated with anticoagulant drugs such as warfarin, they have to be monitored very carefully to make sure that they do not bleed internally, particularly in the brain. The decision whether to give warfarin will depend on many factors, including the patient's age and condition as well as other medication they may be taking.

Lifestyle changes
Making lifestyle changes can often help prevent or treat atherosclerosis. For some people, these changes may be the only treatment needed.

- Eat healthy: to prevent or reduce high blood pressure and high blood cholesterol and to maintain a healthy weight.
- Increase your physical activity: Check with your doctor first to find out how much and what kinds of activity are safe for you. This helps to lose weight, if you're overweight or obese.
- Quit smoking, if you smoke. Avoid exposure to secondhand smoke.
- Reduce stress.

To help slow or reverse atherosclerosis, your doctor may prescribe medicines to help lower your cholesterol or blood pressure or prevent blood clots from forming. If you have severe atherosclerosis, your doctor may recommend one of several procedures or surgeries.

....

Additional information – for general knowledge
Surgery - Coronary artery bypass grafting (CABG)
Coronary artery bypass grafting (CABG) is a type of surgery. Coronary artery bypass grafting consists of attaching an artery or part of a vein to a coronary artery, so that the blood has an alternate route from the aorta to the heart muscle. As a result, the narrowed or blocked area is bypassed. An artery is preferred to a vein because arteries are less likely to become blocked later. In one type of bypass grafting, one of the two internal mammary arteries is cut, and one of the cut ends is attached to a coronary artery beyond the blocked area. The other end of this artery is tied off. If an artery cannot be used or if there is more than one blockage, a section of a vein—usually, from the saphenous vein, which runs from the groin to the ankle—is used. One end of the section (graft) is attached to the aorta, and the other to a coronary artery beyond the blocked area. Sometimes a vein graft is used in addition to the mammary artery graft.

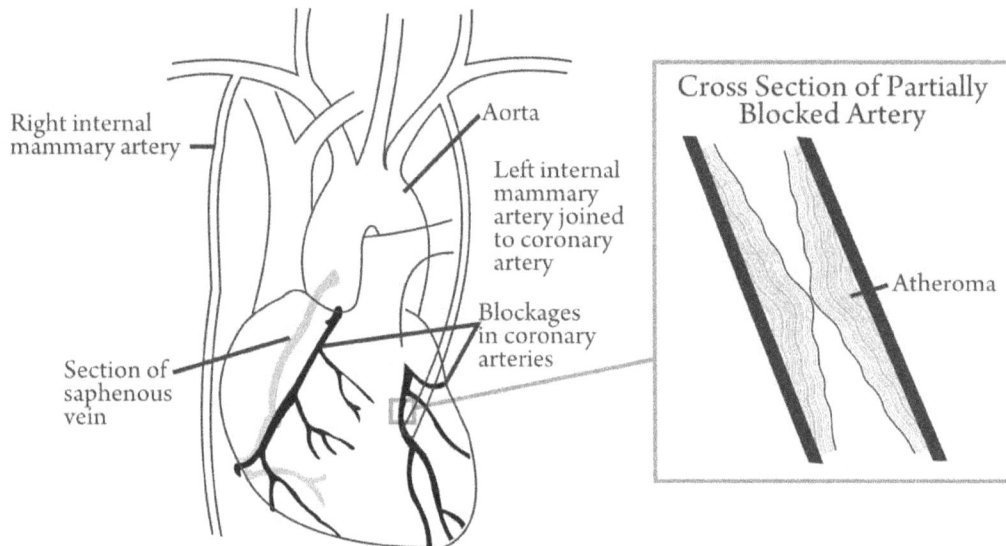

Fig.7.3

Balloon angioplasty

Angioplasty is a procedure to open blocked or narrowed coronary (heart) arteries. Angioplasty can improve blood flow to the heart, relieve chest pain, and possibly prevent a heart attack. Sometimes a small mesh tube called a stent is placed in the artery to keep it open after the procedure.

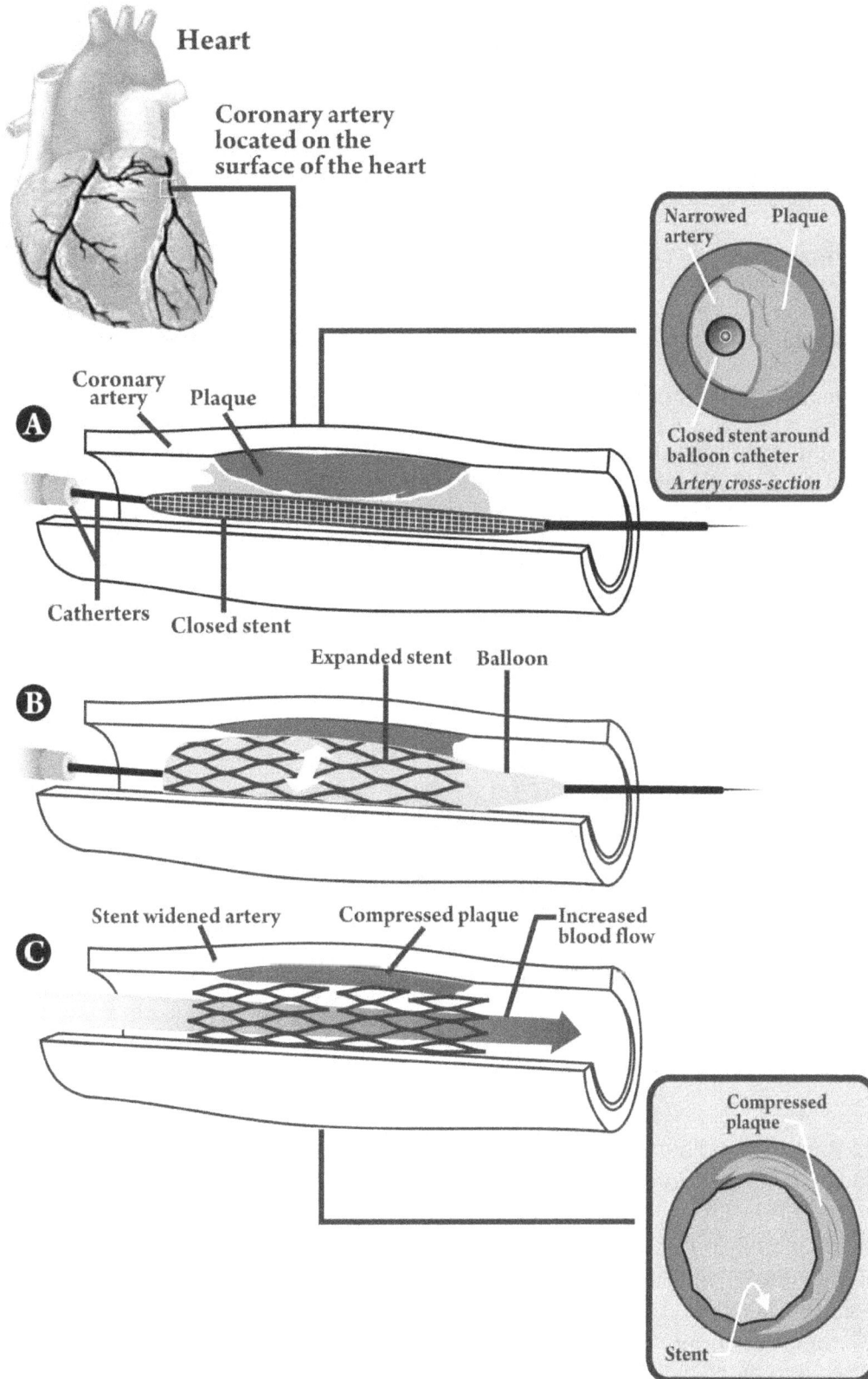

Heart

Coronary artery located on the surface of the heart

Narrowed artery   Plaque

Closed stent around balloon catheter

Artery cross-section

A

Coronary artery   Plaque

Catherters   Closed stent

B

Expanded stent   Balloon

C

Stent widened artery   Compressed plaque   Increased blood flow

Compressed plaque

Stent

Fig.7.4

CHAPTER EIGHT
CHOLESTEROL AND CARDIO-VASCULAR DISEASES

Learning outcomes: by the end of this chapter you should be able to
Edexcel Syllabus Spec 14: *Analyse and interpret data on the possible significance for health of blood cholesterol levels and levels of high-density lipoproteins (HDLs) and low-density lipoproteins (LDLs). Describe the evidence for a causal relationship between blood cholesterol levels (total cholesterol and LDL cholesterol) and CVD.*

Correlation is when a change in one variable is **reflected** by a change in another variable.
Causation is when a change in one variable is **responsible** for a change in another variable.

Two variables may show a correlation, but it does not mean that a causation exists. To establish a causative link, all other **subsidiary variables must be controlled** or taken into account and the study must involve a **large sample size**, have a **control group**, extend for a considerable **duration** and use **statistical analysis** to obtain **reliable data** and draw a **valid conclusion**.

One such study was the **Framingham Heart Study**. The study involved **5,209 men and women** between the ages of 30 and 62 from the town of Framingham, Massachusetts. It began with an extensive **physical examinations** and **lifestyle interviews that were analyzed for common patterns related to CVD development** (subsidiary factors taken into account). Since 1948, the subjects have continued to return to the study every two years for a detailed medical history, physical examination, and laboratory tests, and in 1971, the Study enrolled a second generation - 5,124 of the original participants' adult children and their spouses - to participate in similar examinations. The third generation of subjects was enrolled in 2002.

The results of the study, relating to LDL-C and HDL-C in men aged between 50 to 70 years is shown in fig.8.1.

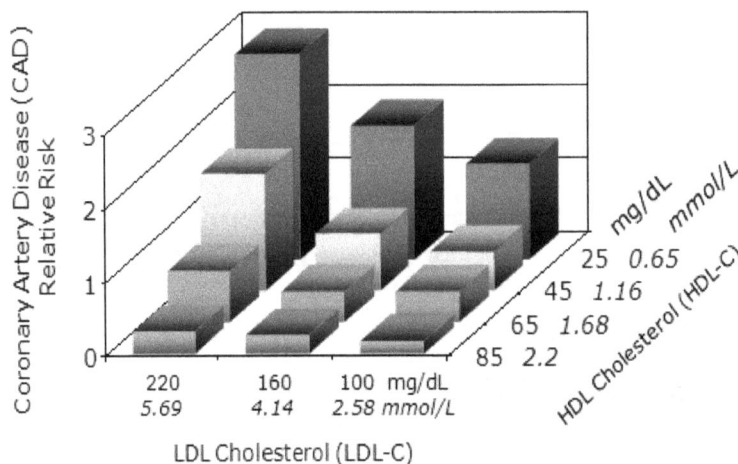

Fig.8.1

The results of the study indicated that:
- The lower the high-density lipoprotein cholesterol (HDL-C) level, the greater is the likelihood of developing coronary artery disease (CAD). A negative correlation.
- This relationship of risk is gradually enhanced as low-density lipoprotein cholesterol (LDL-C) levels are increased. A positive correlation.
- The risk nearly reaches 3-fold when LDL-C is 220mg/dL and HDL-C is 25mg/dL.
- HDL-C remains an important risk factor even in patients with a low LDL-C level. For example, a patient with an LDL-C of 100mg/dL and an HDL-C of 25mg/dL has a similar risk of coronary heart disease as a patient with an LDL-C of 220mg/dL and an HDL-C of 45mg/dL.

Another study, the **Prospective Cardiovascular Münster** (PROCAM) confirms the Framingham data given in the preceding section. PROCAM showed that the incidence of coronary heart disease (CHD) decreased with increasing high-density lipoprotein cholesterol (HDL-C) levels among a group of 4,407 German men, aged 40–65 years, followed for 6 years.

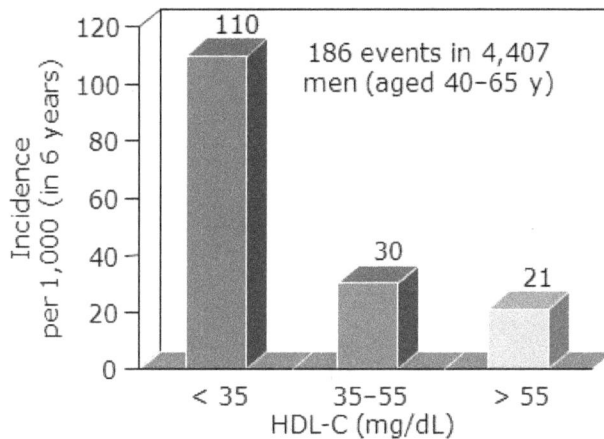

Fig.8.2

A clear and positive relationship between blood cholesterol levels and subsequent coronary heart disease has **repeatedly** (indication of reliability) been demonstrated. Cholesterol in the plasma is transported by lipoproteins (refer to additional material in chapter 6). The cholesterol level associated with the low-density lipoprotein (LDL) fraction is **positively correlated** with coronary heart disease, whereas the cholesterol associated with the high-density lipoprotein (HDL) is **negatively correlated** (the higher the level, the lower the risk), as shown in fig.8.3. These observations have been verified in **several different populations** and have been shown to be **independent of each other, as well as of other known risk factors**. The evidence regarding HDL, although more recent than that for LDL, supports a powerful and independent role for HDL in lowering coronary heart disease risk and probably explains a significant portion of the difference in risk between men and women, with women having higher average levels of HDL than men.

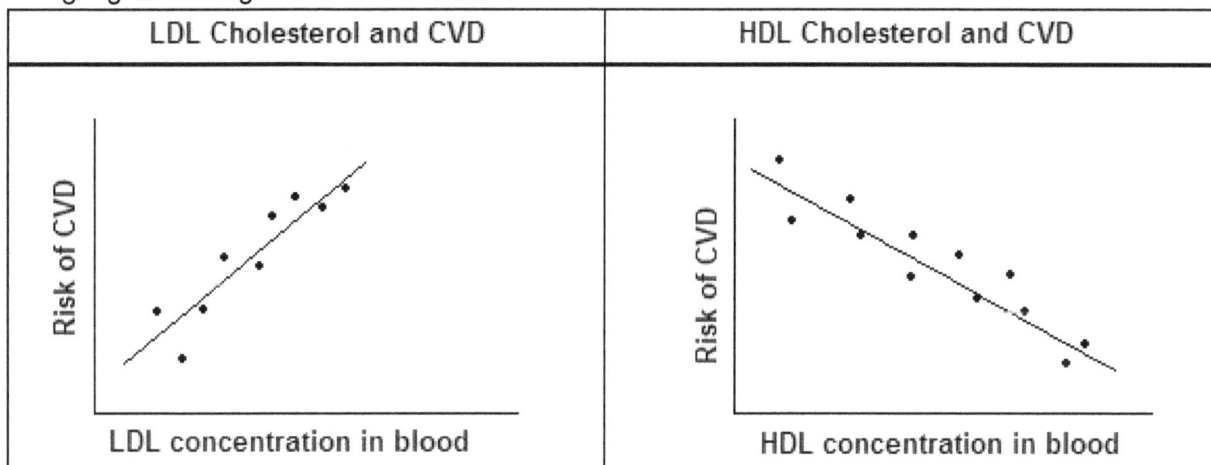

Fig.8.3

A lipid profile is the clinical blood test which measures total cholesterol, LDL cholesterol, HDL cholesterol, and triglycerides. Fig.8.5 shows the classification of cholesterol levels.

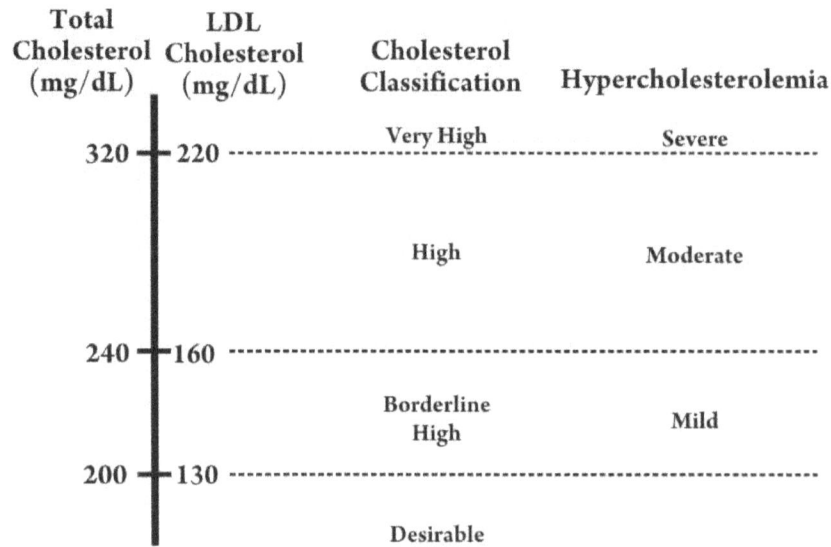

Total Cholesterol (mg/dL)	LDL Cholesterol (mg/dL)	Cholesterol Classification	Hypercholesterolemia
320	220	Very High	Severe
		High	Moderate
240	160		
		Borderline High	Mild
200	130		
		Desirable	

Classification of serum cholesterol concentrations.
100 mg/dL cholesterol = 2.586 mmol/L
Fig.8.5

This evidence from the Münster Study appears to show a clear link between the LDL/HDL ratio and deaths from coronary heart disease. The balance of these Lipoproteins in blood is now recognised as a clear indication of the risk of developing CVDs. A LDL/HDL ratio above 4.0 drastically increases the risk of CVD, as shown in fig.8.6.

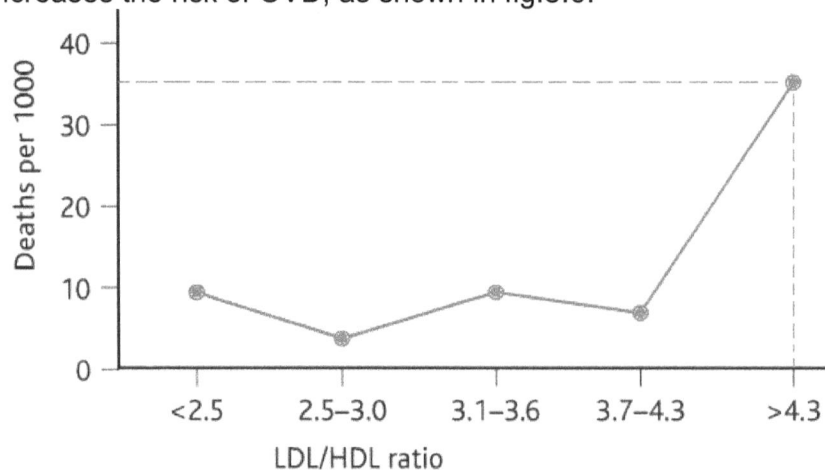

Fig.8.6

Summary
- **The lower the high-density lipoprotein cholesterol (HDL-C) level, the greater is the likelihood of developing coronary artery disease (CAD).**
- **High levels of non–high-density lipoprotein cholesterol (non HDL-C) and low HDL-C concentrations may promote the development of atherosclerosis.**
- **Low HDL-C Levels Increase CHD Risk Even When Total-C Is Normal.**
- **Risk is enhanced as low-density lipoprotein cholesterol (LDL-C) levels are increased.**
- **A LDL/HDL ratio above 4.0 drastically increases the risk of death from CVD.**

CHAPTER NINE
OBESITY

Learning outcomes: by the end of this chapter you should be able to
Edexcel Syllabus Spec 15: Discuss how people use scientific knowledge about the effects of diet (including obesity indicators), exercise and smoking to reduce their risk of coronary heart disease.
Edexcel Syllabus Spec 16: Describe how to investigate the vitamin C content of food and drink.
Edexcel Syllabus Spec 17: Analyse data on energy budgets and diet so as to be able to discuss the consequences of energy imbalance, including weight loss, weight gain, and development of obesity.

Obesity is a risk factor for coronary heart diseases (CHD). Thus one must consume a diet which will prevent the person from getting obese. The energy content of the diet must be balanced out with the physical activity level and Basal Metabolic Rate of the individual. Obesity and risk of CHD can be measured by obesity indicators. The fig.9.1 shows the effects of Obesity.

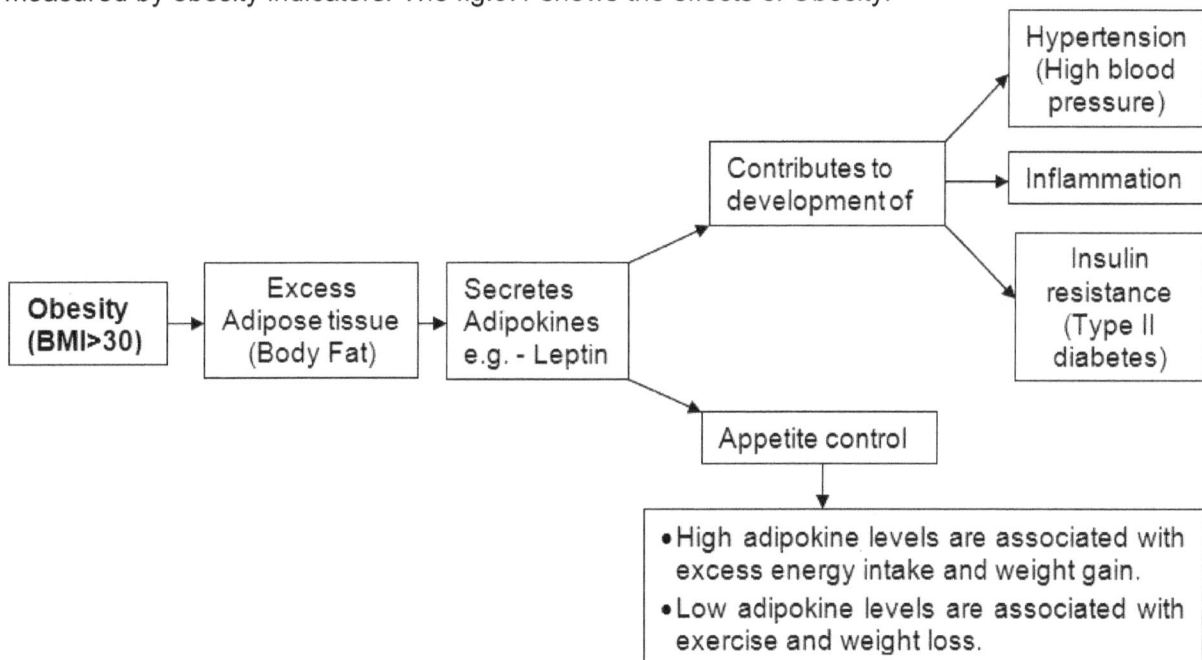

Fig.9.1

Obesity indicators – and objective means of determining obesity

Overweight and obesity increase the risk of CHD. As well as being an independent risk factor, obesity is also a major risk factor for high blood pressure, raised blood cholesterol, diabetes and impaired glucose tolerance'.

The adverse effects of excess weight are more pronounced when fat is concentrated in the abdomen. This is known as **central or abdominal obesity** and is assessed using the waist to hip ratio. The World Health Organization's World Health Report 2002 estimated that over 7% of all disease burden in developed countries was caused by raised body mass index (BMI), and that around a third of CHD and ischaemic stroke and almost 60% of hypertensive disease in developed countries was due to overweight.

More recently the INTERHEART case-control study estimated that 63% of heart attacks in Western Europe and 28% of heart attacks in Central and Eastern Europe were due to abdominal obesity (a high waist to hip ratio), and those with abdominal obesity were at over twice the risk of a heart attack compared to those without. This study also found that abdominal obesity was a much more significant risk factor for heart attack than BMI.

Waist – to - Hip Ratio

The waist to hip ratio is one of the most used clinical applications of girth measurements. This assessment is important *because* there is a correlation between chronic diseases and fat stored in the midsection.

The waist-to-hip ratio can be computed by dividing the waist measurement by the hip measurement by doing the following

1. Measure the *smallest* part of the waist, without drawing the stomach inward, as in fig.9.2.

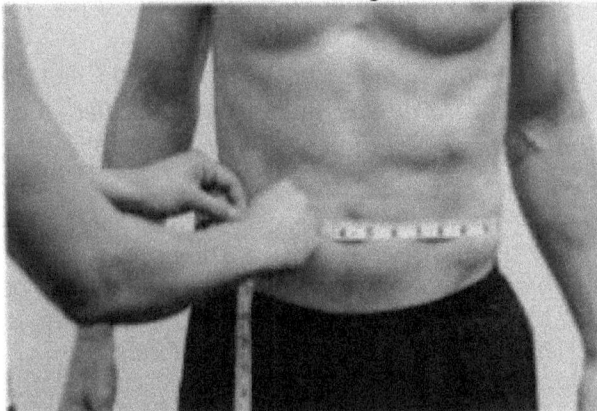

Fig.9.2

2. Measure the largest part of the hip, as in fig.9.3.

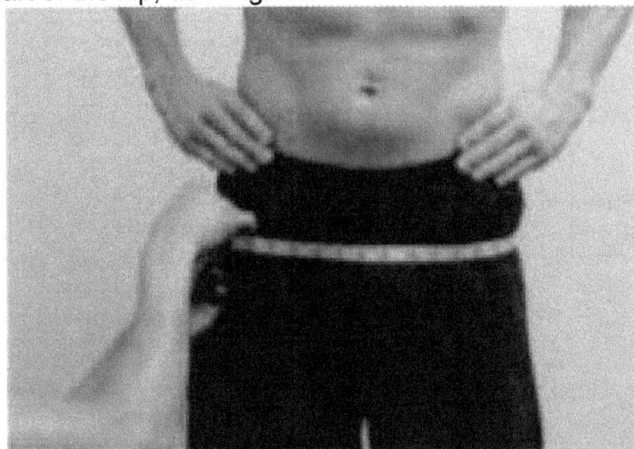

Fig.9.3

3. Compute the hip-to-waist ratio by dividing the waist measurement by the hip measurement.
4. For example. if a client's waist measures 30 inches and his or her hips measure 40 inches, divide 30 by 40 for a hip-to-waist ratio of 0.75. A ratio above 0.80 for women and above 0.95 for men may put these individuals at risk for a number of diseases.

Body Mass Index (BMI)

Although this assessment is not designed to assess body fat, the body mass index (BMI) is a quick and easy method for determining whether a person's weight is appropriate for his or her height. To assess weight relative to height, divide body weight (in kilograms) by height (in meters squared) or, Kg/m^2.

$$BMI = Weight(in\ kg)\ /\ Height^2\ (in\ m)$$

It has been shown that obesity-related health problems increase when a person's BMI exceeds 25. The BMI cut offs listed below apply to 'normal' healthy adults. *It is not appropriate for pregnant women, or for use in some medical conditions or with children. They may also be inappropriate for athletes and some ethnic groups (as BMI does not distinguish between fat and fat-free mass). For example, a BMI over 27.5 in an Asian person has been estimated by the World Health Organisation to carry the same health risk as a BMI of 30 in a white Caucasian person.* It is more difficult to assess obesity in children, so special charts have been developed which take into account growth, gender and age.

This table shows the way different BMIs are classified and the relationship between BMI and risk of associated diseases

Body Mass Index (kg/m^2)	Classification	Risk of disease associated with excess weight
Less than 20	Underweight	Low (but increased risk of other clinical problems)
Over 20 to 25	Desirable or healthy range	Average
Over 25 to 30	Overweight	Increased
Over 30 to 35	Obese (Class I)	Moderate
Over 35 to 40	Obese (Class II)	Severe
Over 40	Morbidly or severely obese (Class III)	Very severe

Fig.9.4

Keep in mind the BMI and hip-to-waist measurement does not take into account percent body fat or lean body mass. Thus, one may possess a high BMI or hip-to-waist ratio score but have a low percentage of body fat and high percentage of lean body mass, indicating their risk for obesity-related health problems may not be as high as their results may show.

Exercise

The easiest way to increase physical activity is by incorporating more physical activity into daily routines, e.g. walking or cycling instead of driving (particularly for short journeys) and taking up more active hobbies such as gardening. The benefits of exercise are shown in fig.9.5.

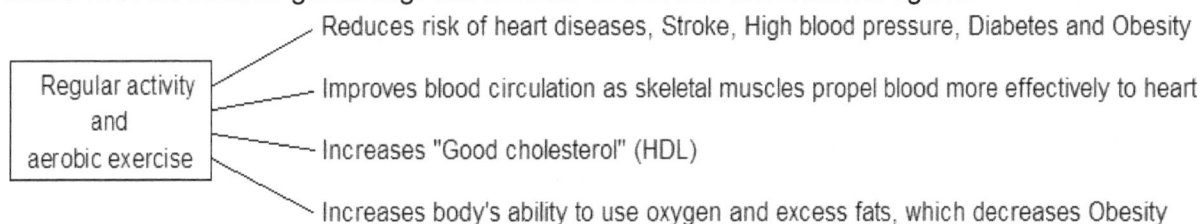

Fig.9.5

Caution: If you have a high risk of heart disease, a history of heart disease, diabetes, or other serious health problem, get your doctor's guidance before beginning or increasing an exercise program.

Smoking

The effects of smoking have already been explained in chapter 6, fig.6.2. The pie chart in fig.9.6 shows the estimated average annual number of smoking-attributable deaths in the United States during 2000 through 2004 by specific causes. The data shows that 32.03% of the deaths are caused by stroke and ischaemic heart disease, associated with cigarette smoking.

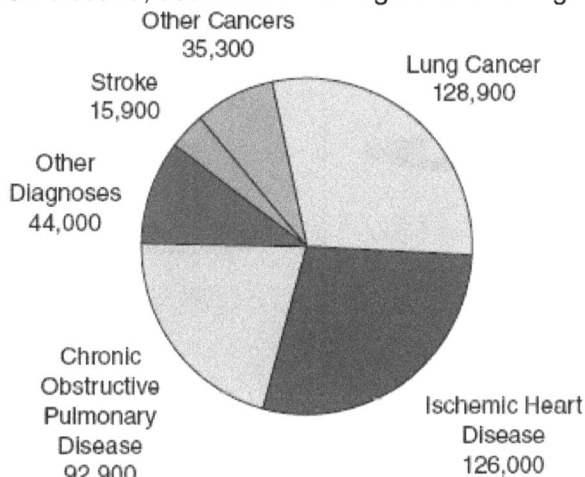

Fig.9.6

Energy basics
- Energy is needed by the body to stay alive, grow, keep warm and be active.
- Energy is provided by food and drink. It comes from the fat, carbohydrate, protein and alcohol. Food and drinks provide different amounts of energy.
- **Different people need different amounts of energy**.
- Some activities use more energy than others.
- **To maintain body weight, it is necessary to expend as much energy as is derived from food; to lose weight, energy must exceed intake**.
- Energy can be measured in either *joules* or *calories*. One calorie is equivalent to 4.2 joules.

How much energy do individuals need?
The actual amount of energy needed will vary from person to person and depends on their *basal metabolic rate* (BMR) and their level of physical activity.

Basal metabolic rate
The basal metabolic rate can be thought of as the rate at which a person uses energy to maintain the basic functions of the body, *i.e.* an adult will use around 4.6 kJ (1.1 kcal) every minute. BMR is measured when a person is at complete rest, and varies from person to person both within a population group and between different groups. Infants and young children have a proportionately high BMR for their size due to their rapid growth and development. Men usually have a higher BMR than women since they tend to have more muscle. Older adults usually have a lower BMR than the young since the amount of muscle tends to decrease with age. The BMR accounts on average for about three quarters of an individual's energy needs.

Energy balance

When the diet provides **more energy** than is needed, the excess is stored as fat and the person **'puts on weight'**. The excess energy is stored in the body as fat in *adipose* tissue, as depicted in fig9.8. When the diet provides **less energy** than is needed, the person **'puts loses weight'**, as depicted in fig.9.7. When the diet provides **the same amount of energy** that is being used, the person's weight **remains constant**, as depicted in fig.9.9.

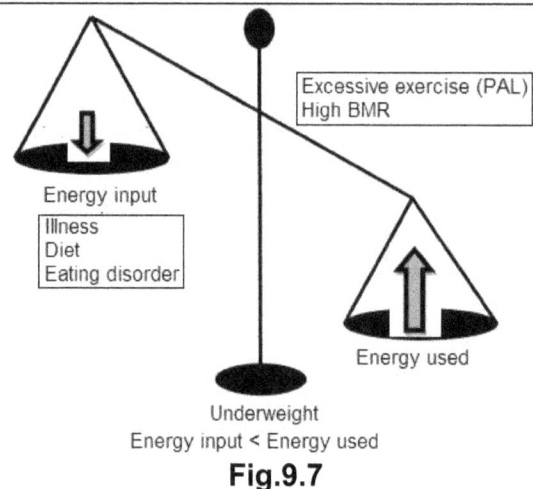

Excessive exercise (PAL)
High BMR

Energy input

Illness
Diet
Eating disorder

Energy used

Underweight
Energy input < Energy used

Fig.9.7

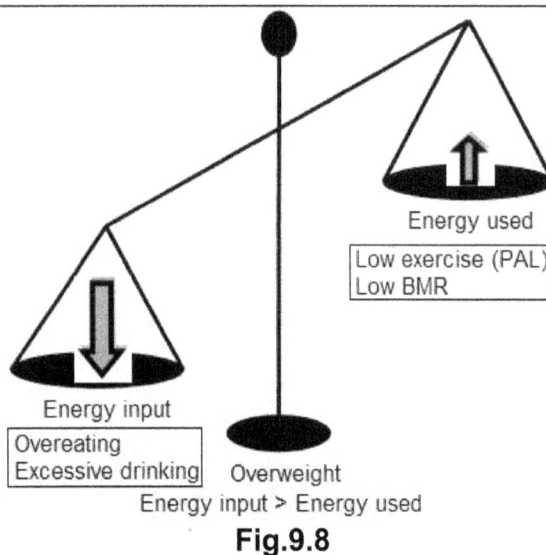

Energy used

Low exercise (PAL)
Low BMR

Energy input

Overeating
Excessive drinking   Overweight
Energy input > Energy used

Fig.9.8

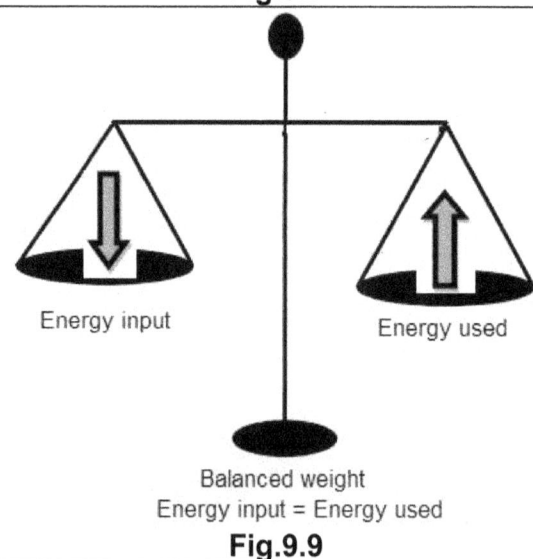

Energy input   Energy used

Balanced weight
Energy input = Energy used

Fig.9.9

The table below shows the energy in a meal and the physical activities that could use up the energy consumed **(Energy in versus energy out)**

Energy consumed = 400 KJ	Physical activity to burn 400 kJ of energy	
White rice boiled (68g) Milk (205ml) Margarine (13g) Milk Chocolate (18g) Wholemeal bread (44g) Tomatoes (548g) Oranges (peeled) (253g)	**Boys**	**Girls**
	Running (16.5 minutes) Walking (30.5 minutes) Cycling (16.5 minutes) Swimming (23.5 minutes)	Running (17.5 minutes) Walking (32.5 minutes) Cycling (17.5 minutes) Swimming (25.0 minutes)

1. Dietary Reference Values (DRVs)

Dietary Reference Values (DRVs) comprise a series of estimates of the **amount of energy (EAR)** and **nutrients (RNI)** needed by different groups of healthy people in the UK population.

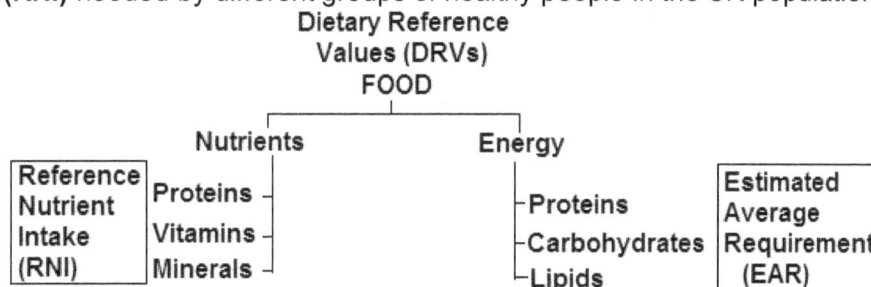

Dietary Reference
Values (DRVs)
FOOD

Nutrients   Energy

Reference
Nutrient   Proteins –
Intake   Vitamins –
(RNI)   Minerals –

–Proteins
–Carbohydrates
–Lipids

Estimated
Average
Requirement
(EAR)

Fig.9.10

a. **Energy:** The **energy recommendations** for diet are provided by Estimated Average Requirement Values **(EAR).** It is estimated by taking into account the physical activity levels (PAL) and the (Basal Metabolic Rate) BMR of the population. The data is then analysed and a reference value is estimated.

Energy EAR = BMR x Physical Activity Level (PAL).

The table below shows the estimated average requirement (EAR) of energy for various groups in the UK population.

EAR - MJ/day (kcal/day)					EAR - MJ/day (kcal/day)				
Age	*Males*		*Females*		*Age*	*Males*		*Females*	
	(MJ)	(kcal)	(MJ)	(kcal)		(MJ)	(kcal)	(MJ)	(kcal)
0-3 mo	2.28	(545)	2.16	(515)	11-14 yr	9.27	(2220)	7.72	(1845)
4-6 mo	2.89	(690)	2.69	(645)	15-18 yr	11.51	(2755)	8.83	(2110)
7-9 mo	3.44	(825)	3.20	(765)	19-50 yr	10.60	(2550)	8.10	(1940)
10-12 mo	3.85	(920)	3.61	(865)	51-59 yr	10.60	(2550)	8.00	(1900)
1-3 yr	5.15	(1230)	4.86	(1165)	60-64 yr	9.93	(2380)	7.99	(1900)
4-6 yr	7.16	(1715)	6.46	(1545)	65-74 yr	9.71	(2330)	7.96	(1900)
7-10 yr	8.24	(1970)	7.28	(1740)	74+ yr	8.77	(2100)	7.61	(1810)

b. Nutrients: the nutrient requirements are provided in the form of Reference Nutrient Intake values (RNIs). **Nutrients** are obtained from **proteins, vitamins and minerals.**

The table below shows the Reference Nutrient Intake (RNI) for Vitamin C for different groups.

Nutrient	Age group	RNI (mg / day)
Vitamin C	0 – 6 months	40
	7 – 12 months	50
	1 – 3 years	15
Also known as Ascorbic acid	4 – 8 years	25
(Dehydro ascorbic acid – DHA)	Males	
	9 – 13 years	45
	14 – 18 years	75
	19 – 30 years	90
(Function: Cofactor for	31 – 50 years	90
reactions requiring reduced	50 – 70 years	90
copper or iron metalloenzyme	> 70 years	90
and as a protective	Females	
antioxidant)	9 – 13 years	45
	14 – 18 years	65
	19 – 30 years	75
	31 – 50 years	75
	50 – 70 years	75
	> 70 years	75

Note: Similar tables of RNI exist for other vitamins, minerals and proteins.

Reference Nutrient Intake (RNI): The RNI is the amount of **a nutrient** that is enough to ensure that the needs of nearly all the individuals in the group (97.5%) are being met. If the average intake of a group is at RNI, then the risk of deficiency in that group is very small.

Lower Reference Nutrient Intake (LRNI): The amount of **a nutrient** that is enough for only the small number of people who have low requirements (2.5%). Intakes below this amount will almost certainly be inadequate, with harmful consequences for health.

How should DRVs be used?

For practical purposes, the RNI should be used when **assessing** the dietary intake of a **group**. The nearer the average intake of the group is to the RNI, the less likely it is that any individual will have an inadequate intake. The nearer the average to the LRNI, the greater the probability that some individuals are not achieving adequate intakes. When planning a diet for **a group** the aim should be to provide the RNI. The fig.9.11 shows the distribution of nutrient requirement within a population.

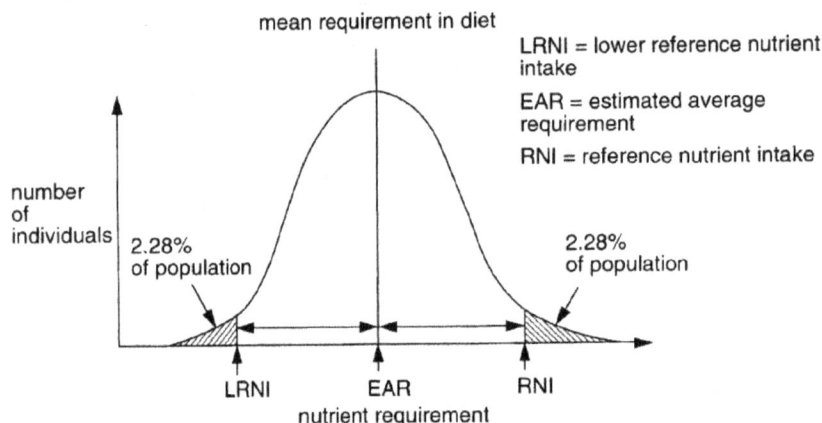

Fig.9.11

The advantage of publishing these values is that it helps to plan diets for individuals or groups of people in the population. It can also be useful for assessing the diets of people. The labels on food products have the calorific values to help people to calculate how much of energy they need to consume to keep healthy.

For example, if you are a 17 year old boy, living in the UK, you would need to consume
- **2755 kcal of energy (EAR) and**
- **75 mg of vitamin C (RNI) every day.**

You could then look at food labels to check for the calorific value of the food you consume. It should ideally be about 2755 kcal per day. If you consume less than the recommended value, then you could be underweight and may suffer from deficiency disorders. If you consume more than the recommended value, you may become obese. Likewise, food labels also give the amount of vitamin C per unit volume of juice. You could calculate how much of a particular fruit juice you would need to consume in order to get the 75 mg of vitamin C per day (RNI).

The EAR for women who become pregnant increases by 0.8 MJ/day (200 kcal/day) but only in the final three months. Although energy is needed for the growth of the foetus and to enable fat to be deposited in the mother's body, pregnant women can compensate for these extra demands by becoming less active and using energy more efficiently.
Breastfeeding mothers have increased requirements for energy but this will depend on the amount of milk produced, the fat stores that have accumulated during pregnancy and the duration of breastfeeding.

Core practical on estimation of Vitamin C concentration of fruit juice.
PRINCIPLE: Vitamin C is an anti-Oxidant, which reduces the blue dye dichlorophenolindolephenol (DCPIP) to a colourless solution (pink colour in the case of some fruits - it is the disappearance of the blue colour, that should look for).

MATERIALS REQUIRED:
Beaker(100 cm^3) , Beaker(500 cm^3), Container to collect a small volume of lemon juice, Measuring cylinder(250 cm^3), Pipette or Syringe to measure 2 cm^3 volume, Pipette or syringe to measure accurately volumes up to 1 cm^3 , Test-tube, Spatula, Distilled water(50 cm^3), Vitamin C (0.1 %, fresh, 0.1 % is 1 mg per cm^3 or lg per L), DCPIP (Dichlorophenolindolephenol, 1 % aqueous solution,freshly made up), Fruits and vegetables.

PROCEDURE :
- Add Vitamin C solution, drop by drop, with a pipette, to 2 cm^3 of the DCPIP solution in a test tube. Shake the tube gently after the addition of each drop and continue to add drops until the DCPIP solution is decolourised. Record the exact volume of vitamin C you added. Repeat the procedure and calculate the mean volume.
- Calculate the mass of vitamin C which is required to decolourise 2 cm^3 of DCPIP solution, knowing that the vitamin C solution was made up to contain 1 mg vitamin C in 1 cm^3 water.
- Repeat the procedure with the fruit juice, containing vitamin C at unknown concentration, instead of the known concentration of vitamin C. If only one or two drops of fruit juice are required to decolourise DCPIP, dilute the juice five times and try again.
- Using the same technique, compare the vitamin C contents of different food and drinks.

CHAPTER TEN
CORRELATION, CAUSATION AND STUDY DESIGN

Learning outcomes: by the end of this chapter you should be able to
Edexcel Syllabus Spec 18: Analyse and interpret quantitative data on illness and mortality rates to determine health risks (including distinguishing between correlation and causation and recognising conflicting evidence).
Edexcel Syllabus Spec 19: Evaluate design of studies used to determine health risk factors (including sample selection and sample size used to collect data that is both valid and reliable).
Edexcel Syllabus Spec 20: Explain why people's perceptions of risks are often different from the actual risks (including underestimating and overestimating the risks due to diet and other lifestyle factors in the development of heart disease).

Correlation and causation
Correlation is when a change in one variable is **reflected** by a significant change in another variable.
Causation is when a change in one variable is **responsible** for a change in another variable.

Any scientific study requires the collection and interpretation of data. It is often necessary to establish whether there is a significant relationship between two variables. This relationship can be **statistically determined by the use of correlation coefficients**. The correlation coefficient can be interpreted to determine whether there is a positive correlation, negative correlation or no correlation at all between the two variables.

However, a correlation does not imply causation.

A positive correlation means that the dependent variable will increase as the independent variable increases, as shown in fig.10.1. For example, the relationship between the amount of annual rainfall and the yield of crops may show a positive correlation. However, it would be inappropriate to conclude that an increase in annual rainfall <u>causes</u> the yield of crops to increase. This is because other factors, like temperature, mineral content of soil, soil type, light intensity and a range of other factors could also have an effect on the yield of crops. However, **if these factors are kept constant** then it may be **more indicative that the volume of annual rainfall causes the crop yield to increase**.

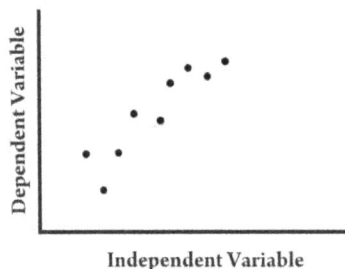

Fig.10.1

A negative correlation implies that the dependent variable will decrease as the independent variable increases. e.g the number of cigarettes smoked per day and the lung capacity.

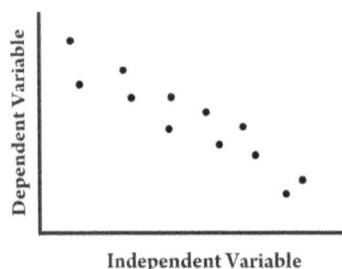

Fig.10.2

A correlation coefficient will simply confirm whether the two variables are correlated or not. It **does not imply any causation**. **Further investigation** needs to be done in order to **identify the causative agent**. If the dependent variable can be influenced by many factors (Multi-factorial) then all other factors (except the one under investigation) must be controlled and the results can then be analysed to determine a causative agent.

Data analysis and conflicting evidence – example One

The table below shows the relationship between blood cholesterol levels and death from coronary heart disease in men in the UK.

Serum cholesterol / mmol dm^{-3}	Frequency in population	Deaths from CHD per 1000 men each year
4.1 – 4.9	22	2
4.9 – 5.7	26	5
5.7 – 6.5	30	8
6.5 – 7.3	12	12
7.3 – 8.2	9	18
8.2 – 8.8	4	25
8.8 – 9.6	4	30

The data clearly shows that as the **serum cholesterol** levels increase, the **death from CHD** also increases. There is **a correlation between the two factors**, however **high serum cholesterol may not be the cause** for the increased incidence of CHD.

In recent years we have learned that it is not the cholesterol in the diet that causes CHDs, but it is the relative levels of HDL and LDL in the blood which is the **cause** of plaque formation and CHDs.

Data analysis – example two

The data on age-specific death rates from coronary heart disease between 1980 and 2005 is shown in the table below.

Death rates per 100 000 of population of the United Kingdom due to coronary heart disease

Age	35 – 44		45 – 54		55 – 64		65 – 74	
	Men	Women	Men	Women	Men	Women	Men	Women
1980	56	9	270	50	733	215	1621	688
1990	37	6	159	33	536	179	1352	594
2000	19	5	92	20	291	84	823	347
2005	19	4	73	16	204	54	558	225

The data can be interpreted to draw the following conclusions.

Conclusion one: The average risk of death due to coronary heart disease increases as the population grows older. This is substantiated by the fact that the death rate increases with age for both men and women, during the entire period of the study. For example, in 1980 the death rate for 35 to 44 year old men is 56 per 100 000 of the population. This number rises in 65 to 74 year old men to 1621 per 100 000 of the population. The same trend is seen for women. This may be due to the fact that arteries get less elastic as the person ages. This increases the blood pressure and increases the risk of heart diseases.

Conclusion two: The number of deaths from coronary heart disease between 1980 and 2005 decreases. For example, the death rate in 45 to 54 year old women decreases from 50 per 100 000 of the population in 1980 to 16 per 100 000 of the population in 2005. A similar trend is seen for other age groups as well.

This may be due to improvements in their diet, which may contain less saturated fat due to increased awareness of risks and lifestyle factors. The awareness factor may also contribute to increased physical activity or exercise to minimise the risk of coronary heart diseases. Improved screening and diagnostic tools help to identify people who are at greater risk and allows them to make lifestyle and diet changes. Drugs like statins, are used as a preventive measure to keep the LDL levels low and reduces the risk of Coronary Heart Diseases (CHD).

Data analysis and conflicting evidence – example Three

The graphs below show the incidence of angina per 100 000 adults determined in two UK studies, the 4th National Study of Morbidity Statistics from General Practice (1991 /1992) and One general practice in Oxford (1989 / 1991).

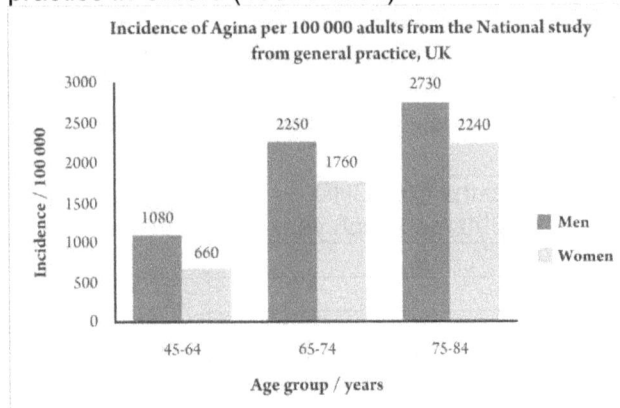

Fig.10.3

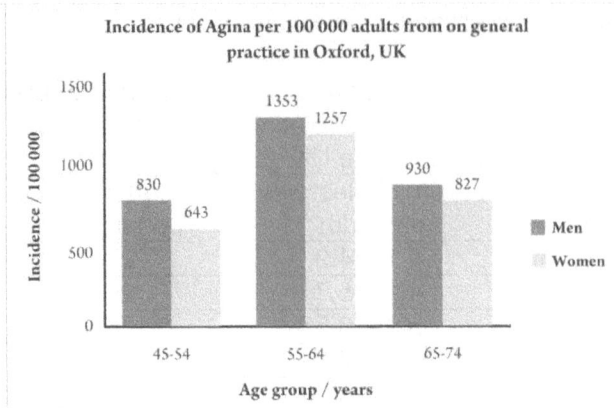

Fig.10.4

The following conclusions can be drawn from the studies.

Conclusion one

At all ages, men have a higher incidence of death from coronary heart disease than do women. The difference between men and women decreases in the older age groups, as shown in fig.10.3 and 10.4.

This may be because men have a greater genetic predisposition. Men may be consuming more alcohol and smoking more than women, which increase the formation of atheroma.

Oestrogen in women has a protective effect on the heart as it reduces the LDL levels in blood. So, women are at lower risk before menopause. After menopause, the protective effect of oestrogen is not present, so the difference in risk is lower in older age groups.

Conflicting evidence

National Study from General Practice (1991 – 1992)		Study of one general practice in Oxford (1989 – 1991)
This study shows that the risk of angina for both men and women increases with age from 45 to 84	but	the Oxford study shows an increase followed by a decrease after the age of 64 for both men and women
Also, males in the National GP study aged **45-64 had a 1 in 93 chance** of having angina, those between **65 and 74 had a 1 in 45 chance** of angina. **How risk is calculated?** Between the age of 45 – 64, 1080 men out of 100,000 suffer angina symptoms, so it means that 1 out of every 92.6 or 93 men suffered the symptoms. 100,000 ÷ 1080 = 92.6 Or (1080 ÷ 100,000) x 100 = 1.08% of the population suffered the symptoms.	but	in the Oxford study men aged **45–64 had a 1 in 46 chance** of having angina, those between **65 and 74 had a 1 in 107 chance** of angina. This pattern is repeated for women too. **Risk calculation.** Between the age of 65 – 74, 930 men out of 100,000 suffer angina symptoms, so it means that 1 out of every 107 men suffered the symptoms. 100,000 ÷ 930 = 107 Or (930 ÷ 100,000) x 100 = 0.93% of the population suffered the symptoms.
The differences in the two studies may be due to the different year of the study, the size of the sample studied (one is a national study, the other draws data from a single general practice) or the methods used in the study.		

Features of a good study

1. Clear aim
A well designed study should include a clearly stated hypothesis or aim. The hypothesis should be testable and the results must be analysed in a valid and reliable way, by the use of statistics.

2. Representative sample

a. Large sample size
The sample size must be large enough to provide **statistical significance**. It should usually represent a large proportion of the population.

b. Randomised selection
The selection of individuals must be done by randomization to reduce **bias (the inclination or tendency to favour one point of view)**. In randomization, the patients are allocated by chance to the different groups. This procedure is performed with random numbers from a calculator or by picking numbers from a hat. Randomization ensures that the patients will be allocated to the different groups in a balanced manner and that possible confounding factors — such as risk factors and genetic variation — will be distributed by chance between the groups. Randomization is intended to maximize homogeneity between the groups and prevent, for example, a specific therapy being reserved for patients with a particularly favourable condition (such as young patients in good physical condition).

c. Control group
There must also be a control group to compare the results and draw valid conclusions.

3. Duration
The study design should be such that the study can be completed in a feasible time span. If the study is too long, then the proportion of individuals who drop out of the study may be too high. This could give rise to complications while analysing the results and reduce the validity of the study.

4. Blinding
Blinding is another suitable method to avoid bias. Blinding means that neither the patient nor the investigator know the treatment being administered. In single blinding, the patient is unaware which treatment he is receiving, while, with double blinding, neither the patient nor the investigator knows which treatment is planned.

Blinding the patient and investigator excludes possible subjective (even subconscious) influences on the evaluation of a specific therapy (e.g. drug administration versus placebo). Thus, double blinding ensures that the patient or therapy groups are both handled and observed in the same manner. The highest possible degree of blinding should always be selected. The control group is usually given a 'fake drug' or 'sugar pill' called as the **placebo**. The case group is given the actual treatment.

5. Valid and reliable results
A reliable method for collection and analysis of data will produce similar results when used by different people at different times for repeated measurements. **All variables that can affect the results must be controlled or adjusted.**

The Munster study on cardiovascular diseases had the following features.

Aim
To identify the causative and contributory agents of Cardio-vascular diseases.

Sample size and composition
The study involved approximately 30,000 volunteers (one third women, two thirds men), ranging between 15 to 65 years. The people were from diverse working environments, ranging from employees of 52 large companies and the public service in Munster and the Northern Ruhr Area of Germany.

Duration
The study extended for a duration of 6 to 8 years. The rate of participation was an average of 60%.

Collection of data
Data was collected from questionnaires, examination and interview by physician. Measurement of blood pressure, resting electrocardiogram, case history and family history, collection of blood sample after 12-hour fast, etc. were also used to obtain data and to account for these factors on the final results.

Valid and reliable results
Since the data was obtained by following standard procedures and the sample size was large, the data should be reliable. The data were analysed statistically and the results were compared with other similar studies to check the validity of the results.

Calculation of actual risk

Risk describes the probability that a particular event will happen. Probability means the chance or likelihood of the event, calculated mathematically.

In 2005, 335969 people died due to Coronary Heart Diseases (CHD). The total UK population at that time was 60 209 408, so we can calculate the average risk in a year of someone in the UK dying from Coronary Heart Diseases (CHD) as:

$$\text{Risk} = (\text{Number of events} / \text{total entities}) \times 100$$
$$\text{Risk} = (335969 / 60\ 209\ 408) \times 100 = 0.55\%$$
Or

Risk may also be expressed as a ratio or probability. For example the risk of people dying from CHD in 2005, in the UK population is 1 in 179, i.e. (335969 in 60 209408).

WHAT IS RISK PERCEPTION?

Risk perception is **one's opinion of the likelihood of risk** (the probability of facing harm) associated with performing a certain activity or choosing a certain lifestyle.

Risk is a normal part of everyone's daily life; there is no such thing as a lifestyle with zero risk. Our perception of risk is often influenced by whether we feel in control of a perceived risk. For example: we may choose to smoke, knowing the risks associated, but may become enraged by a perceived risk that we feel was imposed upon us, for example, living beside a nuclear power plant. Whereas, true risk (a scientifically-evaluated risk) would evaluate the risk of smoking as a higher risk activity. Understanding risk assessment (defining a risk by understanding it) allows you to make an informed decision whereby you weigh the risk of a certain activity with the benefits derived from that activity (a risk versus benefit decision).

The Nature of Risk

We are exposed to risk each day. *Risk can be defined as "the chance you take of becoming injured by a hazard."* Risk measurement starts with probability (odds or chances). What are the odds or chances that we can be injured by a specific hazard? Most people do not judge the probability of risk very well. Risks can be great (leading to death) to negligible (splinter in the finger).

Risk perception (how we judge risk) is an important concept in healthy living. Human perceptions of risk are not very accurate. Our judgments about risks are based upon several things.

People overestimate the risk of something happening if the risk is:

- **Involuntary (Not under their control):** Most people feel safe when they drive. Having the steering wheel in their hands produces a feeling of power, a sense of being in control. If we change places and ride in the passenger seat, we feel nervous because we are no longer in control. When people feel that they have some control over the process that determines the risk facing them, that risk will probably not appear so great as in the case when they have no control over it.

- **Not natural:** Nuclear energy sources, as well as mobile telephones or electric and magnetic fields, are often a greater cause of concern than the radiation produced by the sun. However, it is a well-known fact that the sun is responsible for a large number of skin cancers each year. The natural origin of a risk makes people perceive it as a lesser risk than a man-made one.

- **Dreaded**: Which idea frightens you more, being eaten by a shark or dying of heart disease? Both can kill, but heart problems are much more likely to do so. In spite of this, the most feared deaths are the ones that worry us the most.

- **Unfair**: People who have to face greater risks than others and who do not have access to benefits normally become indignant. The community believes that there should be a fair distribution of benefits and of risks. For example, if the Government builds a nuclear power plant near your house, you will feel at a greater risk than people living far away from the plant.

- ***Possibility of personal impact:*** Any risk can seem greater to us if we ourselves or someone close to us are the victims. This explains why the statistical probability is often irrelevant and ineffective for communicating risks. The closer we are to the risk, and the clearer our knowledge of its consequences, the greater will be our perception of it.

People underestimate the risk of something happening if the risk is:

- **Voluntary:** If your activity is voluntary, then we feel that we are in control of the activity and underestimate the risk.
- **Not immediate (Long term effects – like poor diet):** If the risk is not immediate, then we adopt the 'let's wait and see attitude'. We often feel that we will never be inflicted with the ill effects of the risk factor.
- **Uncertainty or lack of evidence about the effects of the risk:** in every statistical analysis, there is some amount of uncertainty. People usually site the uncertainty to continue indulging in risky activity.

The table below shows the perception of risk by the general public, collected from a survey on risk assessment in the United States. It also shows the actual risk for each factor, which is calculated by experts. The possible reason for overestimation or underestimation is also stated in the last column.

Factor	Experts	Public	Perception by public and reason for estimation
Nuclear Power	1	20	Overestimated – as it is **dreaded** and **not natural**. It is also **Involuntary (Not under their control) and unfair.** The **actual risk** (chance or probability) is a lot lower.
Motor Vehicles	2	1	Underestimated – as it is **voluntary** to drive a motor vehicle and the person feels in **control** of the activity.
Handguns	3	4	Overestimated – as it is **dreaded** by many.
Smoking	4	2	Underestimated – as it is **voluntary** and the **effects are not immediate**.
Fire fighting	5	18	Overestimated – as it is **dreaded** and **not natural**. The **actual risk** (chance or probability) is a lot lower.
Alcoholic Beverages	6	3	Underestimated – as it is **voluntary** and the **effects are not immediate**.
Private Aviation	7	12	Overestimated – as it is **dreaded** and **Involuntary (Not under their control)**. The **actual risk** (chance or probability) is a lot lower.
Police Work	8	17	Overestimated – as it is **dreaded** and **Involuntary (Not under their control)**. The **actual risk** (chance or probability) is a lot lower.
Pesticides	9	8	Underestimated – as it is **voluntary** and the **effects are not immediate**.
Surgery	10	5	Underestimated – as it is **voluntary**.

Scale: 1 represents the lowest risk and 20 represent the highest risk.

Risk perception and development of heart diseases
Diet and other lifestyle factors

It is a well known fact that a diet high in saturated animal fats, sodium chloride and consumption of large amounts of alcohol are factors which increase the risk of developing cardiovascular diseases. Lack of physical activity and smoking are other lifestyle factors that could also increase the risk of suffering from cardiovascular diseases. However, we tend to ignore the risk associated with these components of our diet and lifestyle because of the underestimation or overestimation of the risk.

Additional information – not needed for exams.

In order to find the best possible evidence, it helps to understand the basic designs of research studies. The following basic definitions and examples of clinical research designs are commonly followed. Let us analyse and evaluate the design of the following studies.

1. Case Series and Case Reports: These consist either of collections of reports on the treatment of individual patients with the same condition, or of reports on a single patient. Case series/reports are used to illustrate an aspect of a condition, the treatment or the adverse reaction to treatment.

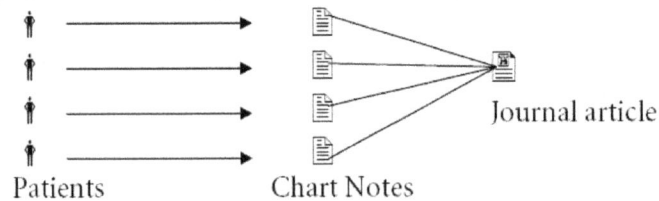

Patients   Chart Notes   Journal article

Fig.10.5

Example: If you need to write up a report on dengue fever, then you could look up database of hospitals (which is usually confidential, but could be accessed with special permission and anonymity of patients). The database usually stores information about the symptoms, diagnostic test results, treatment prescribed and the progress of the patient in response to the treatment. The report could then be written using the data obtained.

Analysis of design

Sample selection: The sample here represents the patients who have visited the hospital for consultation and subsequent treatment. There is no inclusion of patients who have not attended the hospital or taken any treatment for dengue fever.

Sample size: the **sample size will be relatively large** as hospitals store information for long durations. It will however, depend upon how common the disease is in the region.

Reliability: If the results are consistent, which means that we get the same results for many patients from many hospitals, then the results are reliable.

Validity: Because they are **reports of cases** and **do not have a control group** with **which to compare outcomes, they** have low **validity.**

2. Case Control Studies: Patients who already have a certain condition are compared with people who do not. Researchers **look back in time (retrospective)** to identify possible exposures. They often rely on medical records and patient recall for data collection. Case control studies are generally designed to estimate the **risk** of developing the studied condition or disease. They can determine if there is an associational relationship between condition and risk factor.

For example: A study in which colon cancer patients are asked what kinds of food they have eaten in the past and the answers are compared with a selected control group. Case control studies are less reliable than either randomized controlled trials or cohort studies. A major drawback to case control studies is that one cannot directly obtain absolute risk (i.e. incidence) of a bad outcome. The advantages of case control studies are they can be done quickly and are very efficient for conditions or diseases with rare outcomes.

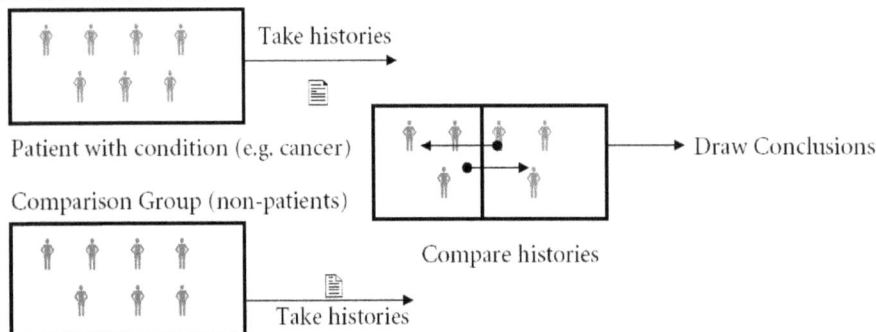

Take histories

Patient with condition (e.g. cancer)

Comparison Group (non-patients)

Compare histories

Draw Conclusions

Take histories

Fig.10.6

Analysis of design

Sample selection: Two groups are generally needed – a control group and an experimental group (case). The control group does not suffer from the condition and the case group suffers from the condition. Other factors, like lifestyle, diet, age, gender, genetics, etc. must be similar in both groups to make the comparison valid and reliable.

Sample size: The larger the size of the sample, the greater is the reliability and validity of the experiment, as it will have greater statistical significance.

Reliability: If the results are consistent, which means that we get the same results for many patients, then the results are reliable. This will happen if the sample selection is done as stated above.

Validity: These types of studies are often less reliable than randomized controlled trials and cohort studies because showing a statistical relationship does not mean that one factor necessarily caused the other (refer to correlation and causation).

3. Cohort Studies or *longitudinal* studies
Cohort studies identify a large population who already has a specific exposure or treatment (case-defined population), follows them over time (prospective) and compares outcomes with another group (control population) that has not been affected by the exposure or treatment being studied.

Cohort studies are observational and not as reliable as randomized controlled studies, since the two groups may differ in ways other than in the variable under study.

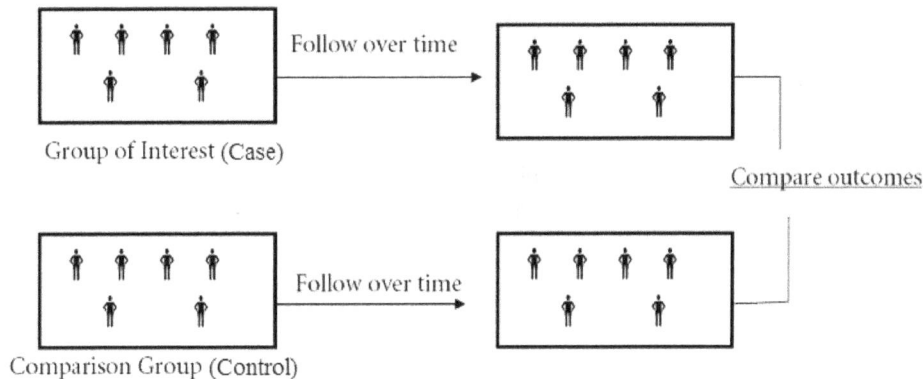

Fig.10.7

Cohort studies may be either prospective (i.e., exposure factors are identified at the beginning of a study and a defined population is followed into the future), or historical / retrospective (i.e., past medical records for the defined population are used to identify exposure factors). Cohort studies are used to establish causation of a disease or to evaluate the outcome or impact of treatment, when randomized controlled clinical trials are not possible.

Example: There may be a group of people with high LDL levels. They may be divided into two groups – a case and a control group. The case group is given statins regularly and the control group does not take statins. The blood LDL levels are then monitored for each group and the results are compared to determine the effect of statins in lowering blood LDL levels.

One of the more well-know examples of a cohort study is the Framingham Heart Study, which followed generations of residents of Framingham, Massachusetts. Cohort studies are not as reliable as randomized controlled studies, since the two groups may differ in ways other than the variable under study. Other problems with cohort studies are that they require a large sample size, are inefficient for rare outcomes, and can take long periods of time.

Analysis of design

Sample selection: It is difficult to find a large group of individuals who have high LDL levels and lead a similar lifestyle, of the same age, gender and genetic makeup. Moreover, the group not taking the treatment will be deliberately put at higher risk of suffering the ill effects of the condition. This is considered unethical and even dangerous. It will require the participants to give an informed **consent**.

Sample size: The larger the sample size the more valid will the results be because variation caused due to other variables will be minimised.

Validity: Cohort studies are observational and not **as** <u>reliable</u> **as** <u>randomized</u> **controlled studies,** <u>since the two groups may</u> differ **in ways other than in the** variable under **study.**

4. Randomized Controlled Studies: a study in which 1). There are two groups, one treatment group and one control group. The treatment group receives the treatment under investigation, and the control group receives either no treatment (placebo) or standard treatment. 2). Patients are randomly assigned to all groups.

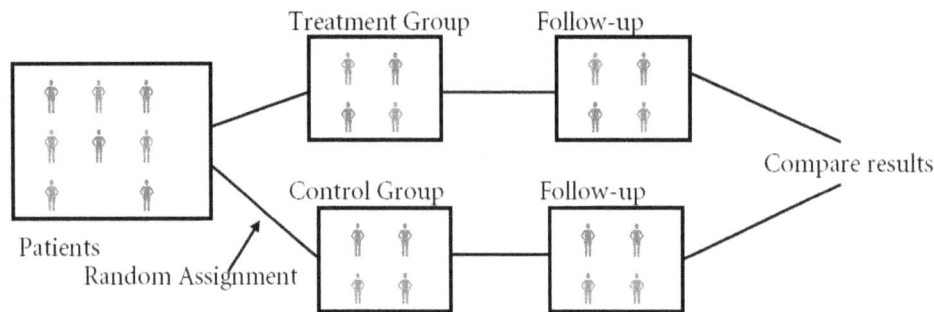

Fig.10.8

Randomized controlled trials are considered the "gold standard" in medical research. They lend themselves best to answering questions about the effectiveness of different therapies or interventions. Randomization helps avoid the bias in choice of patients-to-treatment that a physician might be subject to. It also increases the probability that differences between the groups can be attributed to the treatment(s) under study. Having a control group allows for a comparison of treatments – e.g., treatment A produced favourable results 56% of the time versus treatment B in which only 25% of patients had favourable results. There are certain types of questions where randomized controlled studies cannot be done for ethical reasons, for instance, if patients were asked to undertake harmful experiences (like smoking) or denied any treatment beyond a placebo when there are known effective treatments.

Analysis of design
Sample selection: Selection and grouping of individuals is done on a random basis. This helps to reduce bias.

Reliability: randomisation, blinding and testing in a clinical environment increase the reliability of the results.

Validity: Randomized controlled trials are considered the "gold standard" in medical research. This is because the participants are in a controlled clinical environment. Randomisation of samples and double blinding also improve the validity of the study.

CHAPTER ELEVEN
CELL MEMBRANE STRUCTURE

Models of the Plasma Membrane

Scientists began building models of the cell membranes decades before membranes were first seen with the electron microscope.

An early study of red blood cell membranes revealed that the membrane was composed mainly of **lipids** and **proteins**. The red blood cells were a good choice for the research as they do not have internal membranes or organelles.

Gorter and Grendel model

In 1924 Gorter and Grendel proposed that the cell membrane must actually be **a phospholipid bilayer**. They calculated the **area of the red blood cell membrane** and then extracted the lipids and dissolved it in petroleum ether. The lipids were then allowed to spread into a layer one molecule thick on a surface of water and **the area was measured**.

Total surface area of RBC membrane / mm^2	Surface area of lipid monolayer on water surface/mm^2	Ratio
4.7×10^{-7}	9.4×10^{-7}	1:2

Since, phospholipid molecules have non-polar hydrophobic fatty acid tails and a polar, phosphate hydrophilic head they arrange themselves into a bilayer. The hydrophobic tails faced away from water on the membrane surfaces while the hydrophilic heads were attracted to water on the membrane surfaces, as shown in fig. 11.1.

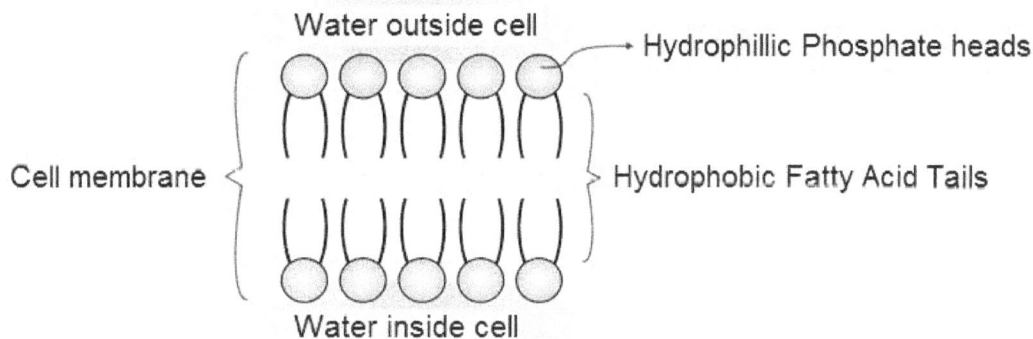

Fig. 11.1

With the conclusion that a phospholipid bilayer was the main fabric of a membrane, the next question was where to place the proteins. It was also clear at that time that a simple phospholipid bilayer was too weak and unstable to hold the cell contents.

Davson and Danielli - sandwich model

In 1935, Davson and Danielli, first proposed the **sandwich model**. It suggested that the stability of the membrane could be accounted for if the **phospholipid bilayer was sandwiched between two layers of proteins.** The first electron micrographs in 1950s supported the Davson-Danielli model.

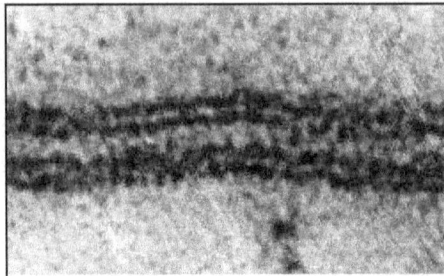

Cell membrane is seen as two dark lines sandwiching a light inner core. Note that this micrograph shows the membranes of TWO adjacent cells.

Fig. 11.2

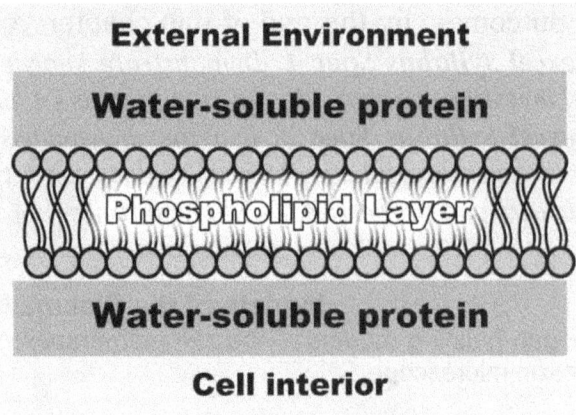

Fig. 11.3

However, many cell biologists in the 1970s recognized **two main problems** with the Davson-Danielli sandwich model.

i. This model assumes that all membranes are identical - this was known to be false:

For example, the cell membrane is 7 to 8 nm in thickness and has a three layered structure in electron micrographs, where as the inner mitochondrial membrane has a thickness of only 6nm and appeared to have bead like structures. Mitochondrial membranes also have a substantially greater percentage of proteins than the plasma membranes.

ii. The membrane proteins would be exposed to hydrophilic environments on all sides (from the phospholipids and from the water of the cytoplasm). This is not a stable configuration:

Analysis of proteins in the membrane revealed that the membrane proteins were amphipathic – having both hydrophilic and hydrophobic regions. This makes the membrane proteins relatively insoluble in water. If such proteins were layered on the surface of the membrane, their hydrophobic regions would be in an aqueous environment.

Singer and Nicolson - Fluid-Mosaic model

Finally, in 1972, S.J. Singer and G. Nicolson proposed the **Fluid-Mosaic model** of the cell membrane. They proposed that the **Cell surface membrane** is made up of a phospholipid bilayer, with **proteins randomly embedded into the phospholipid bilayer,** as shown in fig.11.4.

The fluid mosaic model was able to explain the beaded nature of mitochondrial membranes and could also explain the variable composition of proteins in the membranes.

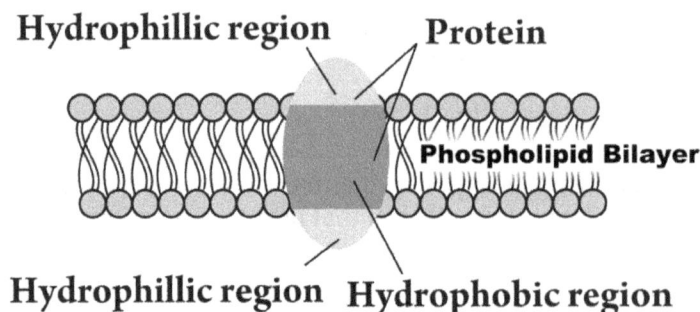

Fig. 11.4

The evidence for this was confirmed by the freeze fracture technique. A cell is frozen and fractured with a knife. The fracture often splits the membrane along its hydrophobic interior. The two layers of phospholipids are split and the proteins are exposed as beads or bumps, demonstrating that the proteins are embedded in the bilayer, as shown in fig. 11.5 and 11.6.

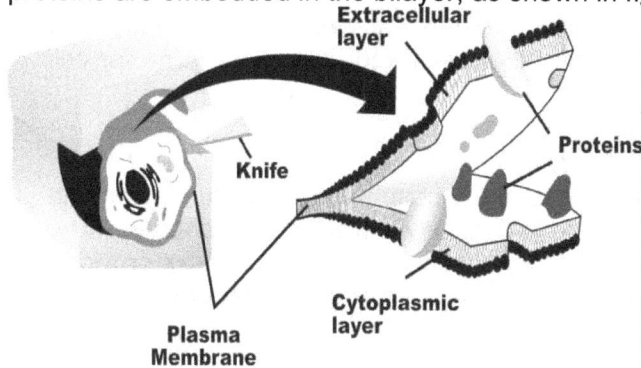

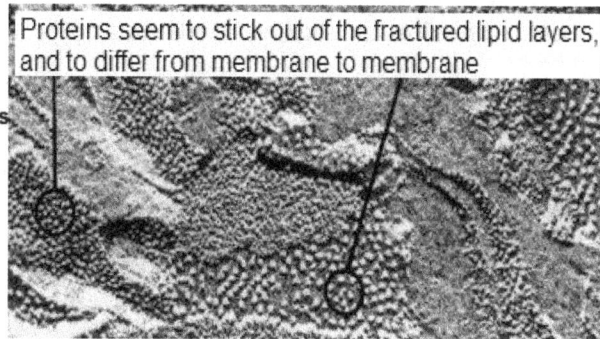

Proteins seem to stick out of the fractured lipid layers, and to differ from membrane to membrane

Fig. 11.5   Fig. 11.6

Meaning of Fluid-Mosaic

Fluid means that **molecules can change places** within the membrane. **Mosaic** means that **proteins are randomly embedded** in to the phospholipid bilayer.

The following experiment shows evidence of the fluid mosaic model.

The plasma membrane proteins of a mouse cell and a human cell were labelled with two different dyes, as shown in fig. 11.7. The two cells were then fused and observed under a microscope. The proteins were found to have changed places and rearranged themselves randomly. This proves that the membrane is a **fluid mosaic**. The phospholipid molecules can also change places laterally within the membrane at the rate of 10 million times a second.

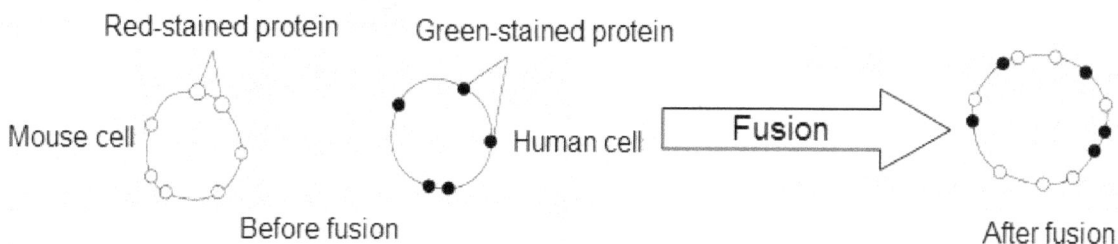

Fig. 11.7

Discovery of aquaporins

It was previously thought that water can diffuse through the phospholipid bilayer because of its small size and low polarity. However, many scientists were not convinced as this could not explain the differential permeability of different cell membranes, specially the cells in the kidney. In 2003, Peter Agre discovered special protein channels called **aquaporins** (Water channels) in the cell membranes. It is now known that water passes through the cell membrane through the phospholipid bilayer and the aquaporins as well. The aquaporins allow the cell to control water movement by opening or closing of aquaporins.

Structure and properties of the cell membrane

The cell membrane is a selectively permeable membrane which helps to prevent passage of some substances and allows the passage of other substances. This helps to maintain the appropriate composition of the cytoplasm.

Phospholipid bilayer

The phospholipid bilayer is **permeable to non polar molecules** like CO_2, O_2 and **lipid soluble molecules** like steroids. The hydrophobic **fatty acid tails form a barrier to polar molecules and ions** like Na^+, K^+, Ca^{2+}, Cl^-, glucose, amino acids, etc.

Proteins

Proteins embedded in the phospholipid bilayer allow **specific** polar molecules and ions to pass across the membrane. Hence, they are often referred to as transporter proteins - **channel proteins and carrier proteins**. Some of the proteins (extrinsic proteins) act as enzymes, recognition sites and electron carriers. Other proteins may serve as **receptors** for hormones and neurotransmitters, play a role in **cell adhesion** by sticking to membrane proteins on other cells and act as **antigens** for cell recognition.

Glycoproteins, glycolipids and cholesterol

Branched chains of carbohydrates maybe attached to some phospholipid molecules, forming **glycolipids** or to proteins, forming **glycoproteins**. This is collectively called the **glycocalyx**. The glycocalyx is always found only on the outer surface of the cell membrane. The glycocalyx provides **chemical protection** to the cell membrane and also plays an important role in **cell recognition and adhesion**.

Cholesterol makes the membranes less fluid and more stable. Unsaturated fatty acid tails in the phospholipid bilayer will increase the fluidity of the membrane.

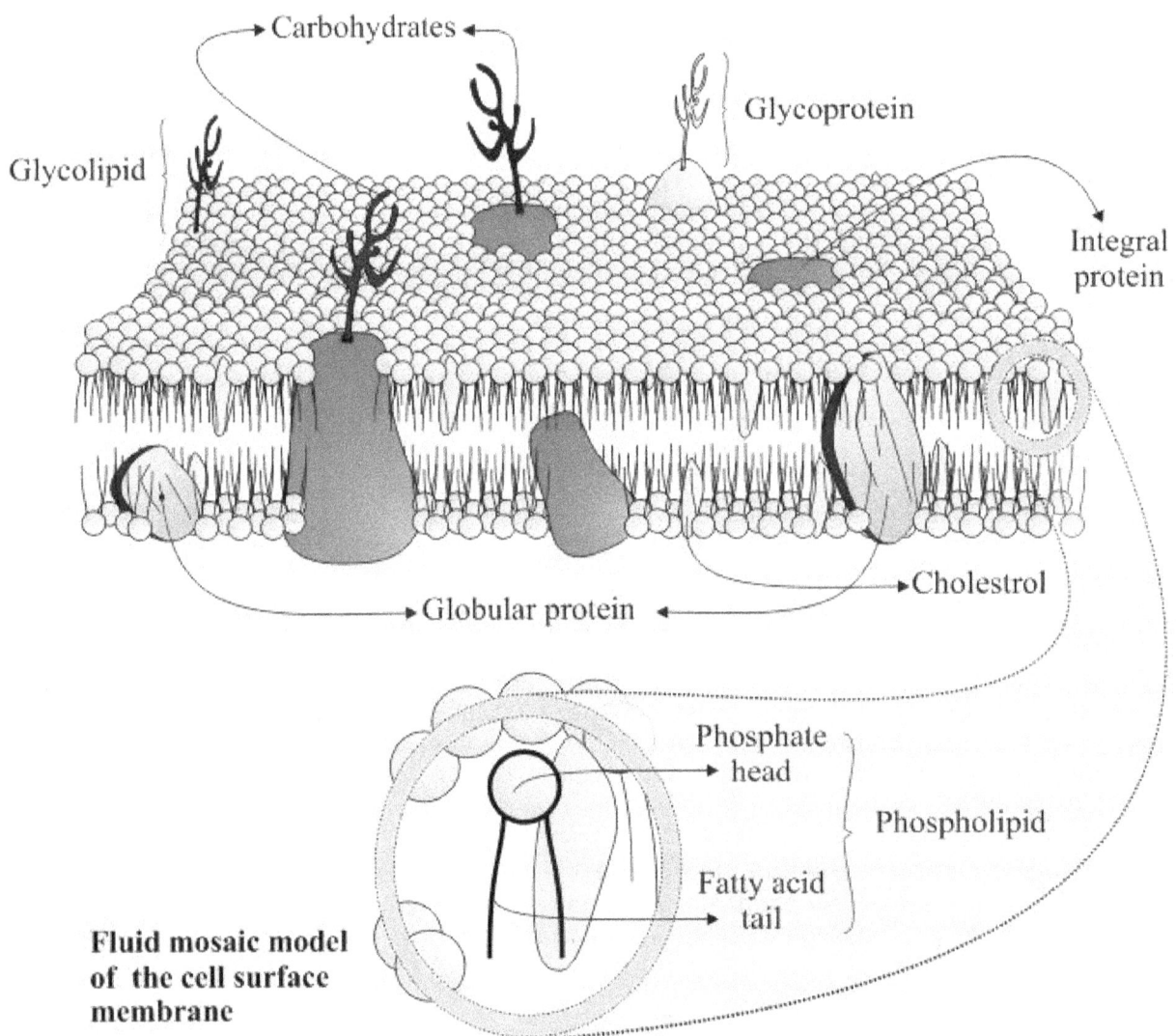

Fig. 11.8

CHAPTER TWELVE
TRANSPORT ACROSS MEMBRANES

Learning outcomes: by the end of this chapter you should be able to
- **Edexcel Syllabus Spec 3:** *Explain what is meant by osmosis in terms of the movement of free water molecules through a partially permeable membrane (consideration of water potential is not required).*
- **Edexcel Syllabus Spec 4:** *Explain what is meant by passive transport (diffusion, facilitated diffusion), active transport (including the role of ATP), endocytosis and exocytosis and describe the involvement of carrier and channel proteins in membrane transport.*

Osmosis – transport of water across the cell membrane

Osmosis is the **net** movement of water molecules from a region of **higher water potential** to a region of **lower water potential** across **a partially permeable membrane**.

or

It may also be described as the **net** movement of **free water molecules** from a region of their **higher concentration** to a region of their **lower concentration** through a **partially permeable membrane**. Consider the diagram below. Water molecules move randomly in both directions across the membrane, but there is net movement from region A to region B.

Explanation: Fig.12.1 illustrates how osmosis occurs. There is **net** movement of water molecules from region A to region B. This is because the solutes in region B attract water molecules. So the concentration of **free water molecules** in region A (**65 free water molecules per unit volume**) is **greater** than the concentration of **free water molecules** in region B (**41 free water molecules per unit volume**). The partially permeable membrane prevents the movement of solute across the membrane.

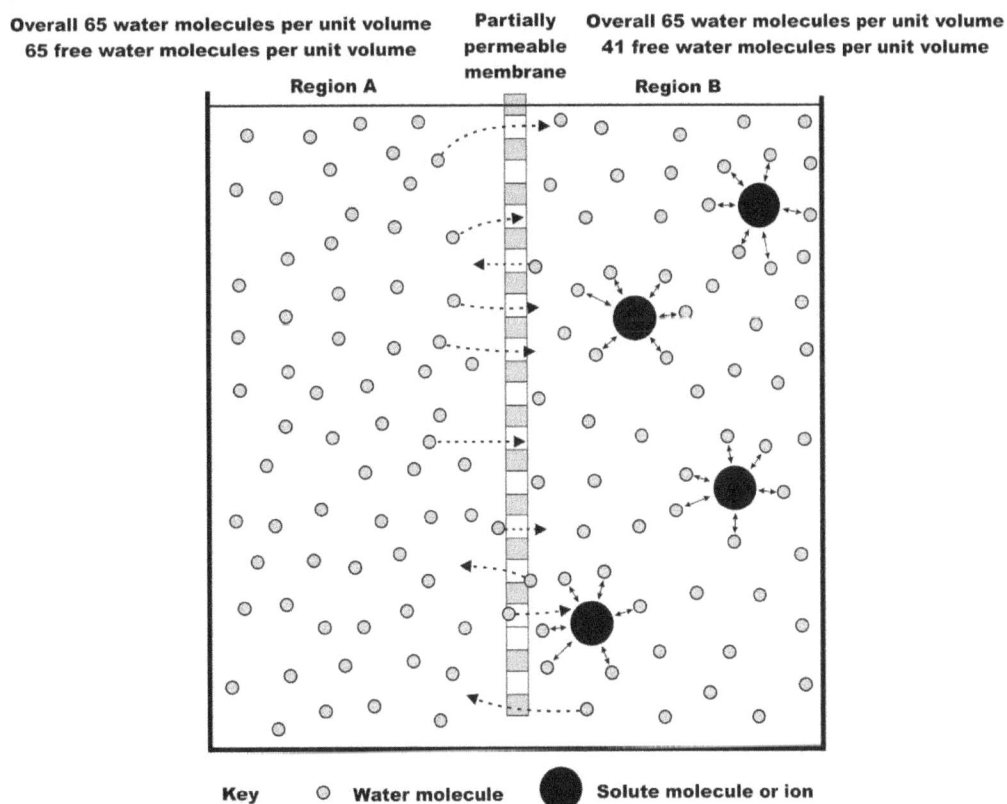

Fig. 12.1

Factors which affect the movement of water molecules during osmosis

Solute concentration

Adding solute into pure water will decrease its water potential. This means that it will reduce the concentration of **free water molecules** in the solution, as the solute molecules will interact with the water molecules. **Insoluble substances** like lipids, starch and glycogen in cells will have **no effect** on osmosis as they do not attract water.

Pressure on the membrane

Even though **free water molecules** may have a tendency to move down their **concentration gradient**, the **movement may be influenced by the pressure on the membrane**. Consider a plant cell placed in pure water. Water enters the cell by osmosis down a concentration gradient of **free water molecules**. However, this does not go on forever. The inward movement of water will stop when the cell becomes turgid, even though a concentration gradient of free water molecules still exists. This is because the cell wall opposes the movement of water molecules into the cells.

Transport of other substances across the cell membrane.

The cell membrane is a selectively permeable membrane which helps to prevent passage of some substances and allows the passage of other substances. This helps to regulate the appropriate composition of the cytoplasm. The Two dimensional structure of the cell surface membrane and its permeability is illustrated in fig. 12.2.

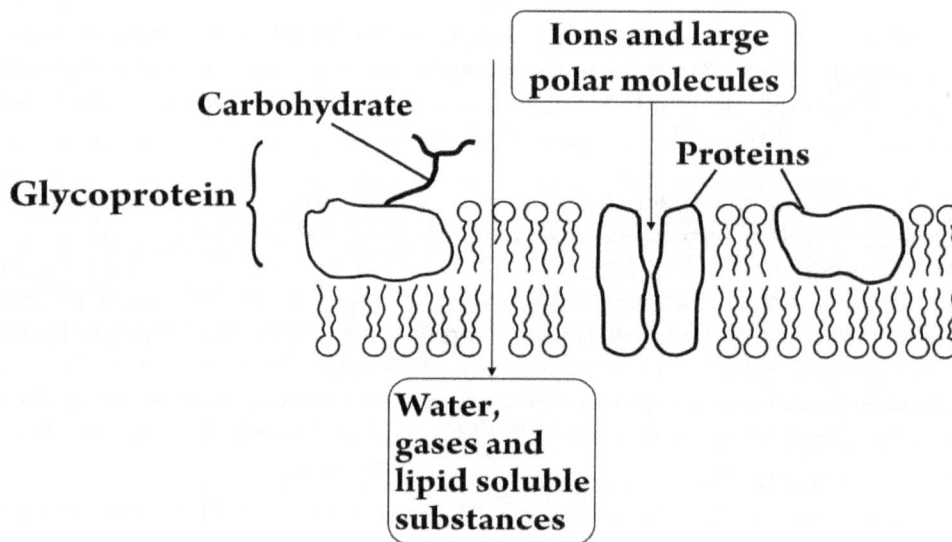

Fig.12.2

Molecules move in and out of cell across the selectively permeable cell membrane by four basic processes, namely, diffusion, facilitated diffusion, active transport and bulk transport (exocytosis and endocytosis).

Simple Diffusion

Diffusion is the **net** movement of particles from a region of its higher concentration to a region of its lower concentration, due to the kinetic energy of the molecules. The particles **move in both directions** across the membrane, but the rate of movement of particles from higher to lower concentration is greater than the movement in the opposite direction. Each type of molecule or ion moves down its own concentration gradient independent of other molecules. For Example: O_2 and CO_2 diffuse in different directions at the same time, in the alveoli.

Substances that can be exchanged by diffusion

- O_2 and CO_2 are non polar, small molecules which can diffuse rapidly across the phospholipid bilayer.
- The non polar fatty acid tails of the phospholipid bilayer serve as a barrier to ions and large polar molecules, like glucose, amino acids, Na^+, Cl^-, etc. So these substances diffuse across the phospholipid bilayer **extremely slowly, if at all**.

- Steroid hormones and glycerol are lipid soluble and can easily diffuse across the phospholipid bilayer of the membrane.

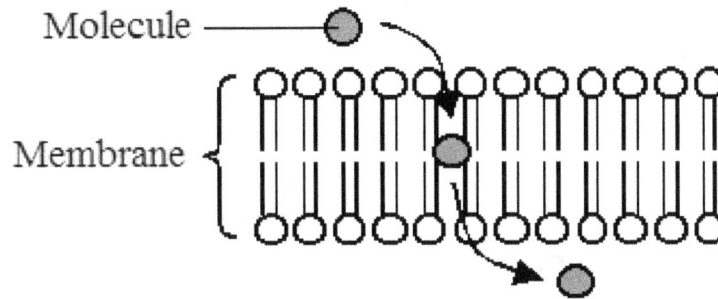

Fig.12.3

Since diffusion is a passive diffusion process, no metabolic energy (ATP) is involved and substances can only move down their concentration gradient. Lipid diffusion cannot be controlled by the cell, in the sense of being switched on or off.

The rate of diffusion across membranes depends on the following factors.
a) Surface area of membrane: the rate of diffusion is directly proportional to surface area of the membrane. The larger the surface area, the greater is the rate of diffusion.
b) Difference in concentration across the membrane: rate of diffusion is directly proportional to the concentration gradient. The larger the concentration gradient, the greater is the rate of diffusion.
c) Thickness of membrane: rate of diffusion is *inversely* proportional to the thickness of the membrane or the diffusion distance. Since cell membranes are extremely thin, the rate of diffusion will be high.
d) Temperature: rate of diffusion is directly proportional to the temperature as the kinetic energy of particles increase with temperature. The higher the temperature, the greater is the rate of diffusion.
e) Size of particles: Smaller and lighter particles diffuse faster, but it will also depend on the permeability of the membrane.

Facilitated Diffusion
Facilitated diffusion is the **net** movement of substances across a cell membrane, **through a trans-membrane protein molecule, without using ATP**. The particles *move in both directions* through the protein, but the rate of movement of particles from higher to lower concentration is greater than the movement in the opposite direction.

The transport proteins tend to be **specific** for one type of molecule, so substances can only cross a membrane if it contains the appropriate protein. As the name suggests, this is a passive diffusion process, so **no metabolic energy (ATP)** is involved and substances can move down their concentration gradient by using **kinetic energy alone**. Some ions (Na^+, Ca^{2+}, K^+, Cl^-, HCO_3^-) and polar molecules (Glucose, amino acids) can diffuse through special transport proteins called **channel proteins or carrier proteins**. Diffusion can occur through the channel in either direction. Since diffusion would not be possible without these proteins the process is called facilitated diffusion. The rate of facilitated diffusion can be limited by the number of transport proteins per unit area of the membrane. Facilitated diffusion is a term which is specific to biology as it can only occur in cells.

There are two kinds of transport proteins that may be involved in facilitated diffusion:

Channel Proteins
Channel proteins form a water-filled pore or channel in the membrane. This allows charged substances to diffuse across cell membranes. Some channels can be **gated** (opened or closed), allowing the cell to control the entry and exit of ions.

Carrier Proteins

Carrier Proteins have a binding site for a specific solute and constantly flip between two states so that the site is alternately open to opposite sides of the membrane. There will be **net** movement of the substance from a high concentration to a low concentration.

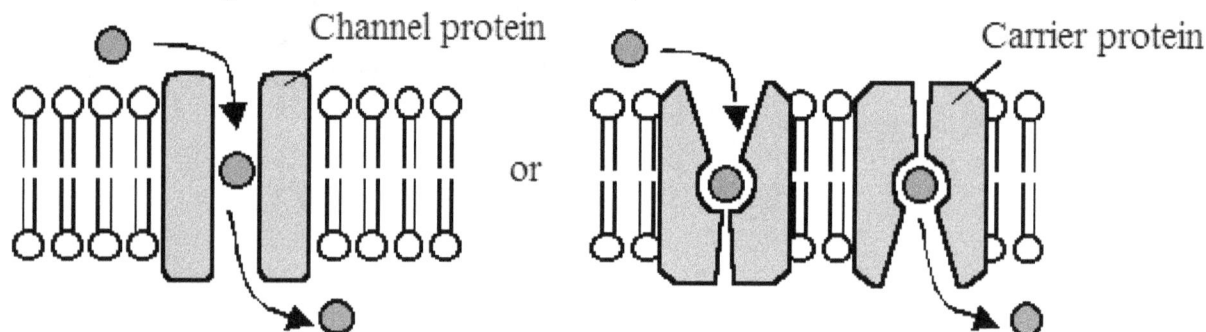

Fig. 12.4

Active Transport

Active transport is the transport of molecules or ions across the cell membrane through **carrier proteins** by using **energy from respiration (ATP).** It transports substances against the concentration gradient.

How active transport works

- The molecule or ion combines with a specific carrier protein in the cell surface membrane.
- ATP transfers a phosphate group to the carrier protein on the **inside of the membrane**. This causes the carrier protein to undergo a change of shape which causes the molecule or ion to move across the membrane.
- The molecule or ion is then released and the protein changes back to its original shape.

Due to energy needed for this process, the cells involved tend to contain **more mitochondria** and have a high rate of respiration. Their ability to carry out active uptake is affected by temperature, oxygen concentration and the presence of respiratory poisons like Cyanide (**all factors which affect the rate of respiration and ATP production**).

Some processes involving active transport are nerve impulse transmission, muscle contraction, absorption of amino acids in ileum, absorption of ions by root hair cells of plants, selective re-absorption in kidney.

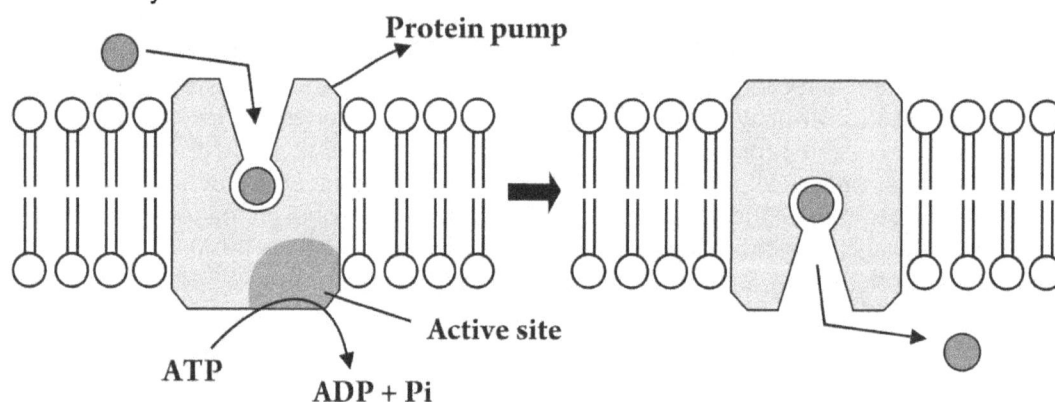

Fig. 12.5

The proteins are highly specific, so there is a different protein pump for each type of molecule to be transported. The protein pumps are also <u>ATPase enzymes</u>, since they catalyse the splitting of ATP into ADP + phosphate (Pi), and use the energy released to change shape and pump the molecule. Pumping is therefore an <u>active process</u>, and is the commonly used transport mechanism that can transport substances <u>up</u> their concentration gradient.

Cytosis – transport involving Vesicles

The processes described so far only apply to small molecules. Large molecules, such as proteins and polysaccharides and even whole cells are moved in and out of cells by using vesicles. ATP is used in the movement of vesicles within the cells.

Endocytosis

Endocytosis is the transport of materials **into** a cell by forming vesicles. Materials are enclosed by a fold of the cell membrane, which then deepens and pinches of from the cell membrane to form a closed vesicle. If the materials are large **solids,** such as a white blood cell ingesting a bacterial cell or an amoeba engulfing bacteria, the process is known as **phagocytosis (cell eating).** These vesicles are relatively large and are called as phagosomes.

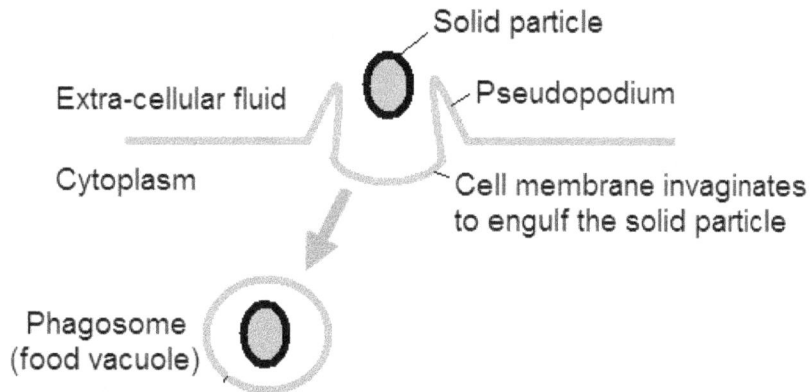

Fig. 12.6

When the materials being taken into vesicles are **liquids or solutions,** such as a protein molecule and the process is known as **pinocytosis (cell drinking).** These vesicles are relatively smaller.

Exocytosis

Exocytosis is the transport of materials **out of a cell by using vesicles**. It is the exact reverse of endocytosis. Materials to be secreted must first be enclosed in a membrane vesicle, usually from the Rough Endoplasmic Reticulum and Golgi bodies. The vesicles move towards the cell membrane and fuse with the cell membrane releasing their contents into the extracellular fluid. Hormones and digestive enzymes are secreted by exocytosis from the secretory cells of the intestine and endocrine glands.

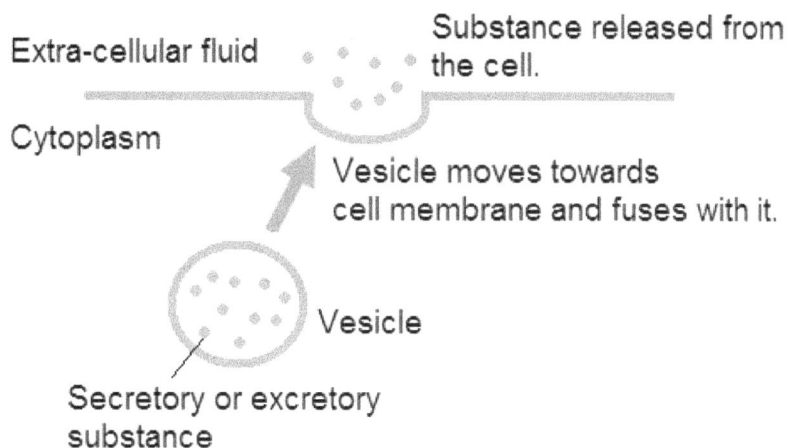

Fig. 12.7

Summary of transport across membranes

METHOD	USES ENERGY	USES PROTEINS	SPECIFIC	CONTROLLABLE
Simple Diffusion	N	N	N	N
Osmosis	N	N	Y	N
Facilitated Diffusion	N	Y	Y	Y
Active Transport	Y	Y	Y	Y
Vesicles	Y	N	Y	Y

Core practical on membrane permeability

Learning outcome: *Edexcel Syllabus Spec 5: Describe how membrane structure can be investigated practically, eg by the effect of alcohol concentration or temperature on membrane permeability.*

Aim: To investigate the effect of temperature and alcohol on cell membrane structure.

PRINCIPLE: Higher temperatures denature the membrane proteins and make the membrane more permeable to the red pigment. So, pigment diffuses out more rapidly.

MATERIALS REQUIRED: Beaker, cork borer, beet root slices, distilled water, water bath, thermometer, colorimeter, cuvette and filter paper.

PROCEDURE :
1. Cut cylinders of beet root tissue from a fresh beet root by using a cork borer. Cut the cylinder into slices of uniform thickness by using a bread slicer. This standardizes the size of beetroot discs and makes surface area to volume ratio constant.
2. Wash the slices with distilled water and blot dry. This will remove dead cells, debris and pigment from the surface of the tissue.
3. Pipette $10cm^3$ of distilled water into a boiling tube and put the five beet root discs.
4. Place the boiling tube in a water bath for 10 min. at $20^0 C$.
5. Repeat the procedure at various temperatures ($30^0 C$, $40^0 C$, $50^0 C$, $60^0 C$, $70^0 C$ and $80^0 C$).
6. Mix the solution and pipette it into a cuvette.
7. Read the absorbance of light using a colorimeter (use blue filter). The absorbance is directly proportional to the extent of membrane disintegration.
8. Tabulate the results and plot graph.

Note: The same experiment can be repeated with different concentrations of ethanol in the boiling tube, kept at a constant temperature, to determine the effect of ethanol on the membrane structure.

CHAPTER THIRTEEN
GAS EXCHANGE IN LIVING ORGANISMS

All living organisms need to respire. Respiration requires oxygen and releases carbon dioxide. If the oxygen supply is less than the demand, anaerobic respiration begins to form lactic acid. Lactic acid accumulates in the cells and decreases the pH of the cytoplasm. This low pH will denature the enzymes, carrier and channel proteins in the cell and the cell will malfunction. So oxygen must be continuously supplied to the cells and carbon dioxide must be removed. Cells also require ATP for active transport, protein synthesis, DNA replication and other metabolic activity. ATP is produced by respiration. So, a continuous supply of oxygen must be provided for the cells to remain metabolically active.

Gas exchange in single celled organisms
In single celled organisms like amoeba, oxygen and carbon dioxide are exchanged directly between the environment and the surface of the organism. This method of exchange, by diffusion alone, is sufficient to sustain the metabolic activity of amoeba because of the following reasons
1. Amoeba has a **large surface area to volume ratio** because of its small size
2. A very **small diffusion distance**. This means that all the surfaces are very near to the centre. So gases can rapidly diffuse from all around the cell to the centre.
3. Respiration constantly uses up oxygen and produces carbon dioxide. This ensures that a **concentration gradient is maintained continuously** and oxygen can diffuse into the cell while carbon dioxide diffuses out.

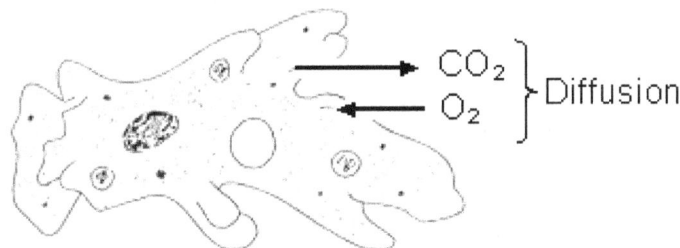

Fig. 13.1

Gas exchange in larger, multi-cellular organisms
Larger organisms like plant and animals have a low surface area to volume ratio and large diffusion distance. Oxygen and carbon dioxide cannot be **rapidly transported** across the body surface. These organisms have developed special respiratory surfaces, like leaves, lungs and gills. The respiratory surfaces have three features in common.
1. **large surface area**
2. **short diffusion distance**
3. **A mechanism to maximize the concentration gradient.**

The relationship between these factors and the rate of diffusion is given by Fick's law, as shown below

$$\text{Rate of Diffusion} \propto \frac{\text{surface area} \times \text{concentration difference}}{\text{distance}}$$

Gas exchange in plants

Plant cells respire all the time, and when illuminated plant cells containing chloroplasts also photosynthesise, so plants also need to exchange gases. The main gas exchange surfaces in plants are the **mesophyll cells** in the leaves. Leaves of course have a huge surface area, and the irregular-shaped, loosely-packed mesophyll cells **increase the area** for gas exchange still further. Moreover, the mesophyll cells have large vacuoles, which push the cell organelles and cytoplasm to the periphery of the cell. This **decreases the diffusion distance** for gas exchange.

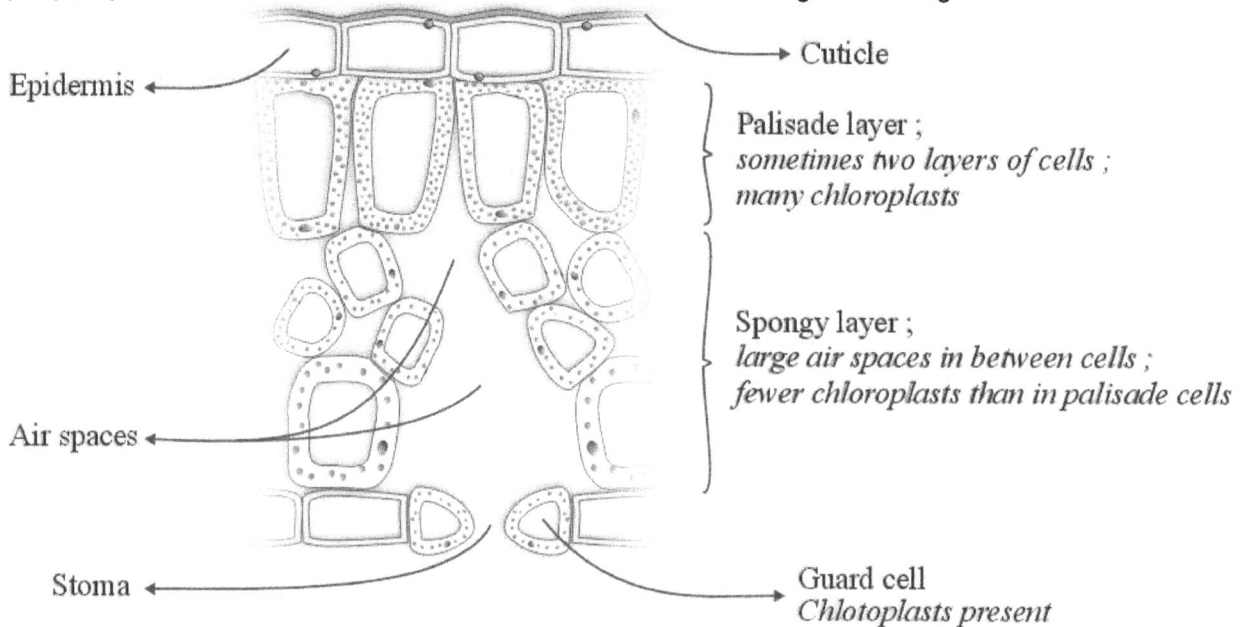

Epidermis

Cuticle

Palisade layer ;
sometimes two layers of cells ;
many chloroplasts

Spongy layer ;
large air spaces in between cells ;
fewer chloroplasts than in palisade cells

Air spaces

Stoma

Guard cell
Chlotoplasts present

Fig.13.2

During the day, photosynthesis increases the oxygen concentration in the sub-stomatal air space, and decreases the carbon dioxide concentration. This increases the concentration gradients for these gases, increasing the rate of diffusion. The exchange of gases between the leaf and the atmosphere occurs through the stomata. This further increases the **concentration gradient**.

The human lungs

The human lungs are composed of millions of tiny sacs, alveoli, dispersed into a matrix of connective tissue called elastin. The elastin matrix also contains numerous blood capillaries, as depicted in fig.13.3.

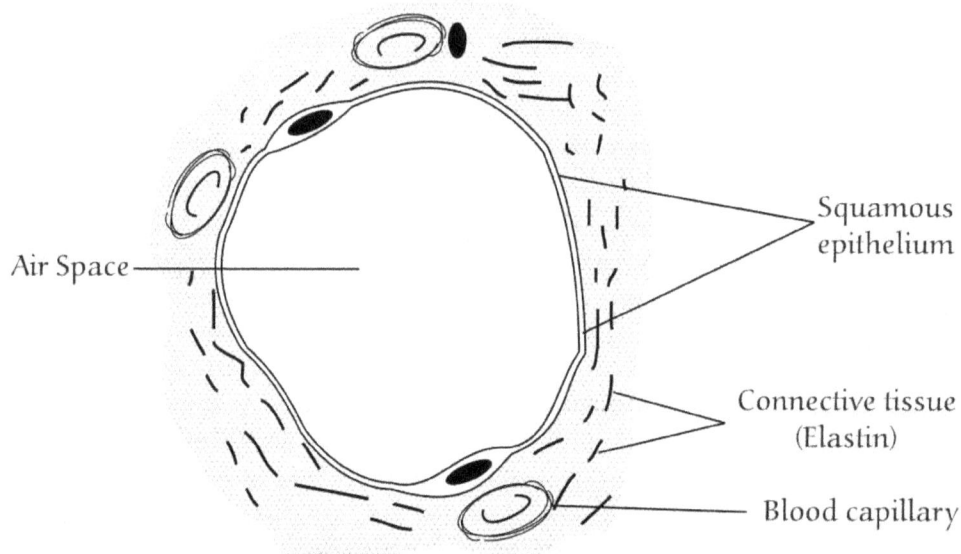

Air Space

Squamous
epithelium

Connective tissue
(Elastin)

Blood capillary

Fig.13.3

The surfactant is a monolayer of phospholipid molecules present on the inner surface of the alveolus as shown in fig.13.4. The lung **surfactant prevents the alveoli from collapsing due to cohesive forces of water molecules during exhalation.** This makes inflation of the alveoli easy. Some babies born without surfactant in the lungs die from exhaustion. This is because they need to spend too much energy to inflate the lungs. This is called the respiratory distress syndrome. The surfactant also **reduces the surface tension of water on the inner lining of the alveoli,** making it easier for gases to pass across the walls.

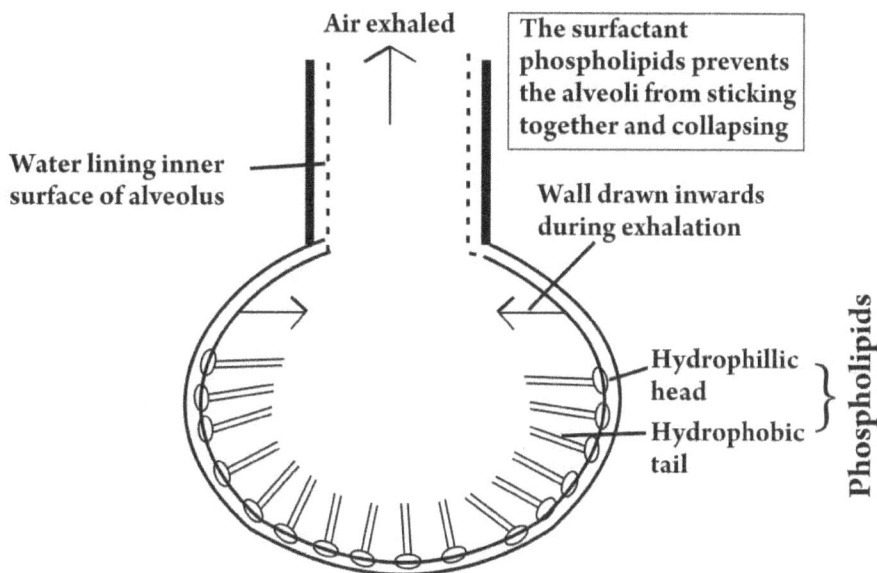

Fig. 13.4

In humans the gas exchange organ system is the respiratory or <u>breathing system</u>. The main features are:

- The walls of the alveoli are composed of a single layer of flattened epithelial cells, as are the walls of the capillaries, so gases need to diffuse through just two thin cells. This decreases the **diffusion distance**.
- The **steep concentration gradient** across the respiratory surface is maintained in two ways: by blood flow on one side and by air flow on the other side.
- There are millions of alveoli and blood capillaries, which provide **an extremely large surface** for gases to be exchanged.
- The flow of air in and out of the alveoli is called **ventilation** and has two stages: **inspiration** (or inhalation) and **expiration** (or exhalation). Lungs are not muscular and cannot ventilate themselves, but instead the whole **thorax** moves and changes size, due to the action of two sets of muscles: the **intercostal muscles** and the **diaphragm**.

During inspiration / inhalation	During exhalation / expiration
• External intercostal muscles contract.	• Internal intercostal muscles contract.
• Internal intercostal muscles relax.	• External intercostal muscles relax.
• Ribs and sternum pulled outwards/upwards.	• Ribs and sternum pulled inwards.
• Diaphragm contracts, becomes flattened.	• Diaphragm becomes dome shaped.
• Volume of thoracic cavity increases causing pressure to become less than atmospheric pressure, so that air rushes into lungs.	• Volume of thoracic cavity decreases.
	• Pressure in lungs becomes greater than atmospheric pressure, so air is pushed out of the lungs.

CHAPTER FOURTEEN
STRUCTURE OF PROTEINS

Learning outcomes: by the end of this chapter you should be able to
- **Edexcel Syllabus Spec 7:** *Describe the basic structure of an amino acid (structures of specific amino acids are not required) and the formation of polypeptides and proteins (as amino acid monomers linked by peptide bonds in condensation reactions) and explain the significance of a protein's primary structure in determining its three-dimensional structure and properties (globular and fibrous proteins and types of bonds involved in three dimensional structure).*

Polypeptides and proteins are chains of amino acids linked to each other by peptide bonds. All the proteins that are needed by humans can be made by the sequential arrangement of 20 amino acids.

Structure of an amino acid

The general structure of an amino acid is shown in fig. 14.1. The amino ($-NH_2$) group and acid ($-COOH$) groups are involved in the formation of **peptide bonds,** which hold amino acids together to form a polypeptide chain.. **Residual (R) groups** are involved in hydrogen bond, ionic bond and covalent bond formation, which **cause the polypeptide chain to fold**.

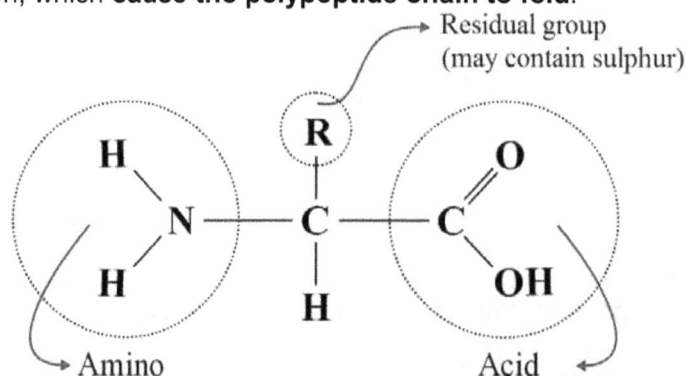

Fig. 14.1

Amino acids are linked to each other by the formation of **a peptide bond,** as shown in fig.14.2. The bond forms by a **condensation reaction** between the acid ($-COOH$) group of one amino acid and the amino ($-NH_2$) group of another amino acid. Two amino acids linked by a peptide bond is called a **dipeptide**, while many amino acids linked by peptide bonds forms a **polypeptide**.

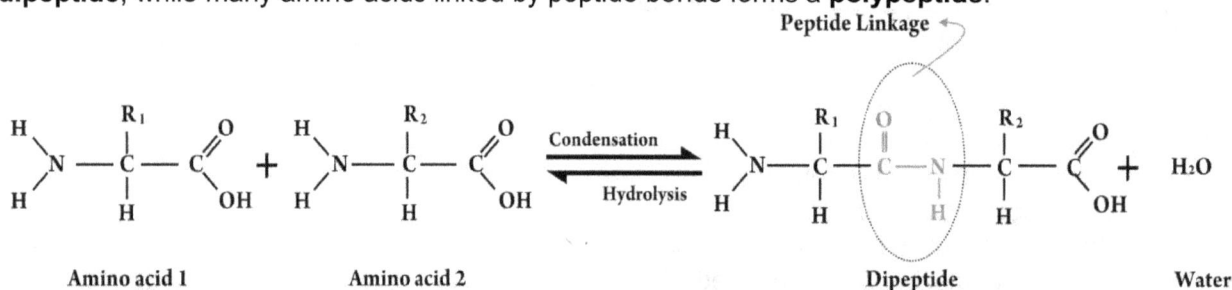

fig 14.2

Condensation is the joining of two amino acids by the removal of a water molecule to form a peptide bond. **Hydrolysis** is the splitting of a dipeptide or polypeptide by the addition of water molecules to break peptide bonds. Both condensation and hydrolysis are brought about by the action of enzymes.

Primary structure of a protein

The primary structure is the **sequence of amino acids** in a polypeptide chain. This sequence is determined by the genetic code on DNA. Finding the primary structure of a protein is called **protein sequencing**. The primary structure determines the secondary and tertiary structure of a protein.
Examples:

$$NH_2 — Leucine — Valine — Isoleunine — Proline — Glutamine — Serine — Arginine — COOH$$

$$COOH — Proline — Valine — Cystine — Proline — Glutamine — Arginine — Leucine — NH_2$$

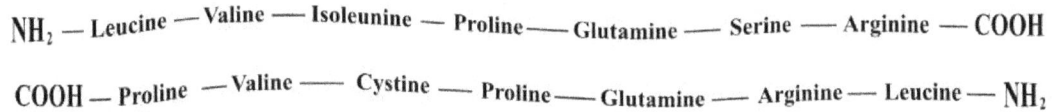

The eventual shape and function of both polypeptide chains is going to be different, as the primary structure is different. However, if two polypeptides have the same primary structure then their secondary and tertiary structure will be similar because they will form similar bo1nds between the amino acids.

Secondary structure of a protein

The secondary structure refers to the folding of the polypeptide chain (primary structure) into helices and pleated sheets. The secondary structure is held together **by hydrogen bonds between the carboxyl groups residues (-C=O) and the amino group residues (-NH) in the polypeptide backbone**. The two secondary structures are the Alpha (α) helix and the Beta pleated (β) pleated sheet.

The alpha(α) helix. The polypeptide chain is wound to form a helix. It is held together by hydrogen bonds running parallel with the long helical axis. There are so many hydrogen bonds that this is a very stable and strong structure. Helices are common structures throughout biology. Example: Keratin in hair, Nails and Skin.

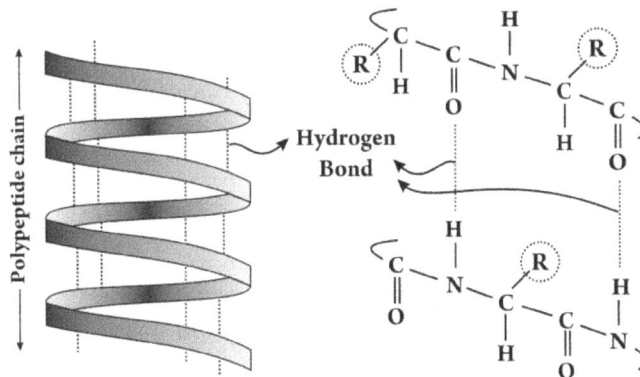

Fig. 14.3 - alpha(α) helix

The β pleated - sheet. The polypeptide chain zig-zags back and forward forming a sheet. Once again it is held together by hydrogen bonds. Example: Fibroin, in silk.

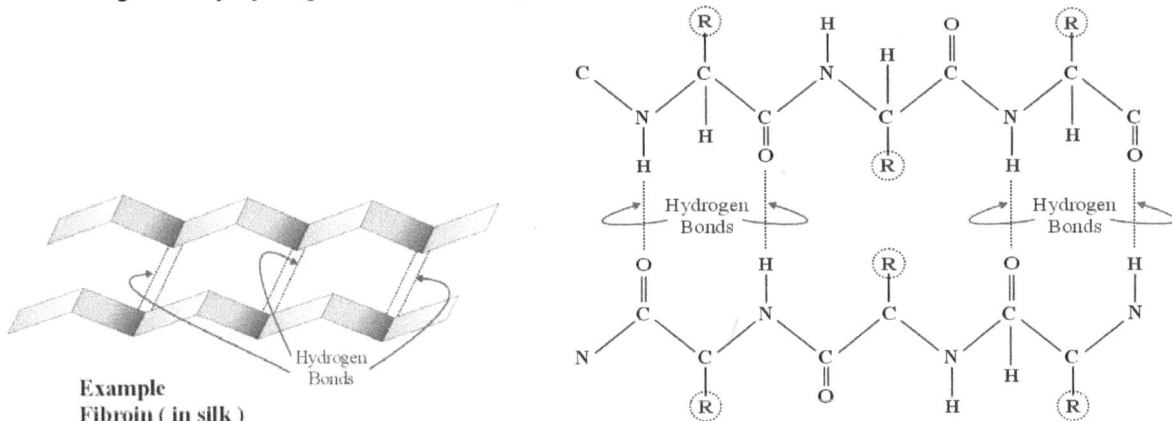

Fig. 14.4 – beta (β) pleated sheet

Tertiary structure of a protein

The tertiary structure of a protein is the complex three - dimensional **globular** shape the polypeptide chain takes when the polypeptide chain twists and folds around itself. The tertiary structure is maintained by Hydrogen bonds, disulphide bridges [covalent bonds] and ionic bonds between the **Residual groups of amino acids**. Hydrophobic interactions also help to maintain the shape of globular proteins, like enzymes and trans-membrane proteins.

The specific three dimensional shape [secondary, tertiary, quaternary structure] of a protein is maintained by three types of chemical bonds between the Residual groups of amino acids

1. **Hydrogen bonds:** form between some Hydrogen atoms [which bear a slight positive charge] and oxygen and nitrogen atoms [which bear a slight negative charge]. Although these bonds are weak, the large number of bonds provide a considerable force to maintain the three dimensional shape.
2. **Ionic bonds:** form between carboxyl [COOH] groups and amino [NH_2] groups found in the Residual chains. They are stronger than H bonds, but can be broken by changes in pH and high temperature
3. **Disulphide bonds:** Some amino acids, like cysteine and methionine contain sulphur atoms in the Residual groups. Disulphide bonds can form between sulphur atoms of amino acids that are close together. These bonds are strong and contribute to the strength of structural proteins like collagen. They are also useful in linking the two polypeptide chains of insulin together.

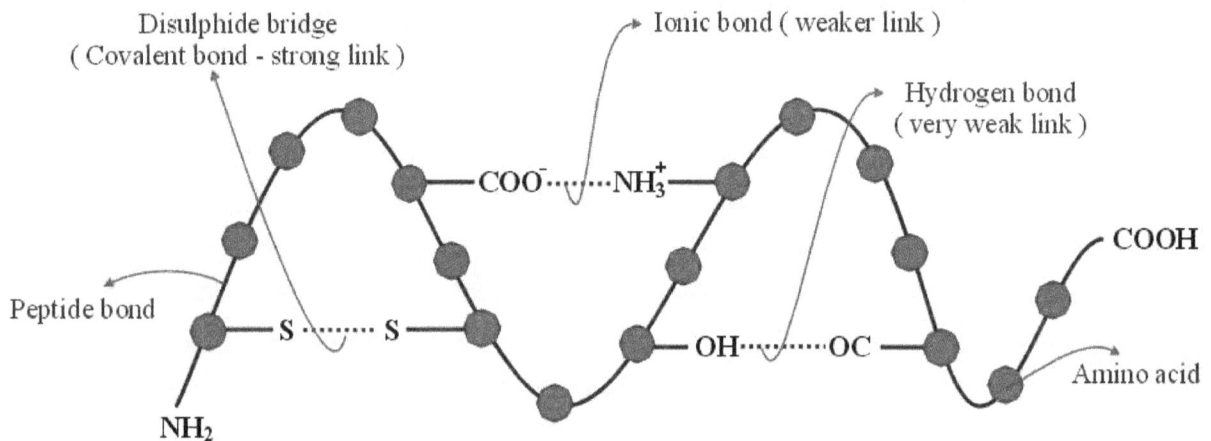

Fig.14.5

Hydrophilic and hydrophobic interactions

Hydrophilic and hydrophobic interactions also help to maintain the shape of globular proteins [tertiary structure] in aqueous solutions. The hydrophobic [water hating] regions of the polypeptide chain face away from water by folding inwards. The hydrophilic regions of the chain remain on the surface of the globular structure.

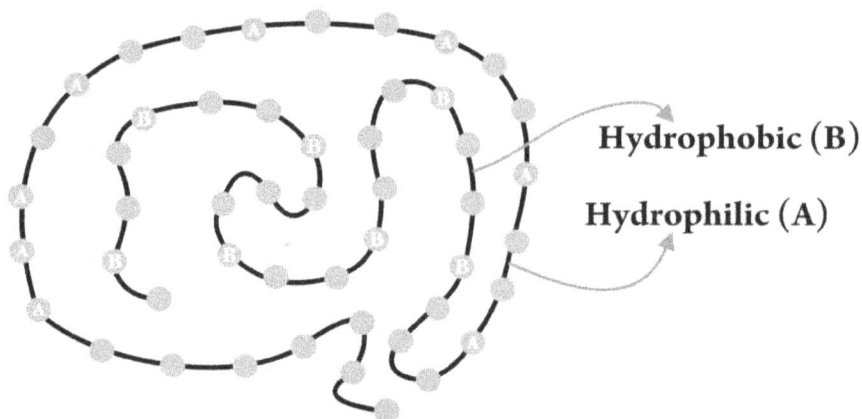

Fig.14.6

Quaternary structure of a protein

Quaternary structure is the linking together of two or more polypeptide chains.

Insulin is a globular protein having a tertiary structure. It also has a Quaternary structure because it is made up of two polypeptide chains which are linked to each other by two disulphide bridges (bonds). The polypeptide chains are highly twisted and rolled up in to a globule when dissolved in water [hydrophobic interactions].

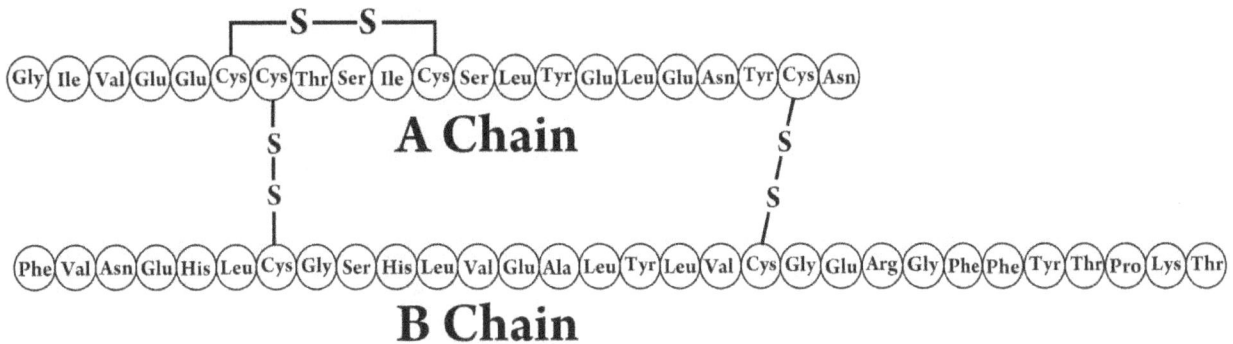

Fig.14.7

Collagen is a fibrous protein. It has a quaternary structure, since it is made up of three polypeptide chains. Each polypeptide chain is twisted to form a helix. The three polypeptide helices wind around each other like a rope with three strands. Hydrogen bonds hold the three strands in place. This makes collagen very stable, insoluble, flexible, but inelastic. Collagen is found mainly in bones and tendons.

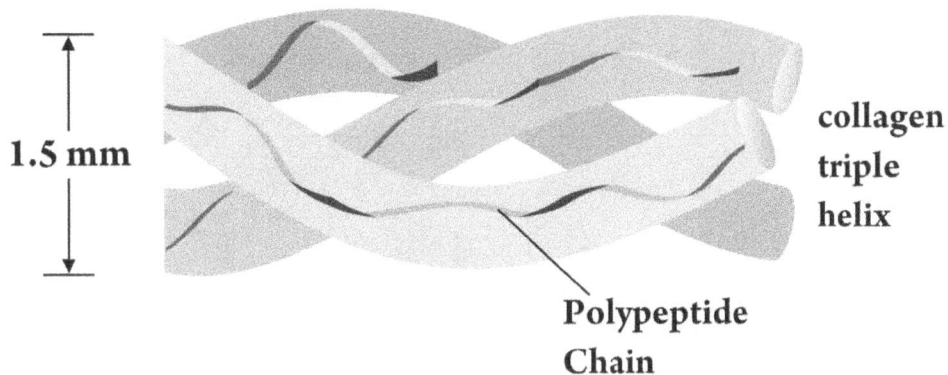

Fig.14.8

CHAPTER FIFTEEN
ENZYMES

Enzyme structure

All enzymes are **globular proteins** - the polypeptide chain is folded into a spherical or globular shape. Hydrogen bonds, ionic bonds, disulphide bridges and hydrophobic interactions (between R groups of amino acids) maintain the specific three dimensional shape of the enzyme, as shown in fig.15.1.

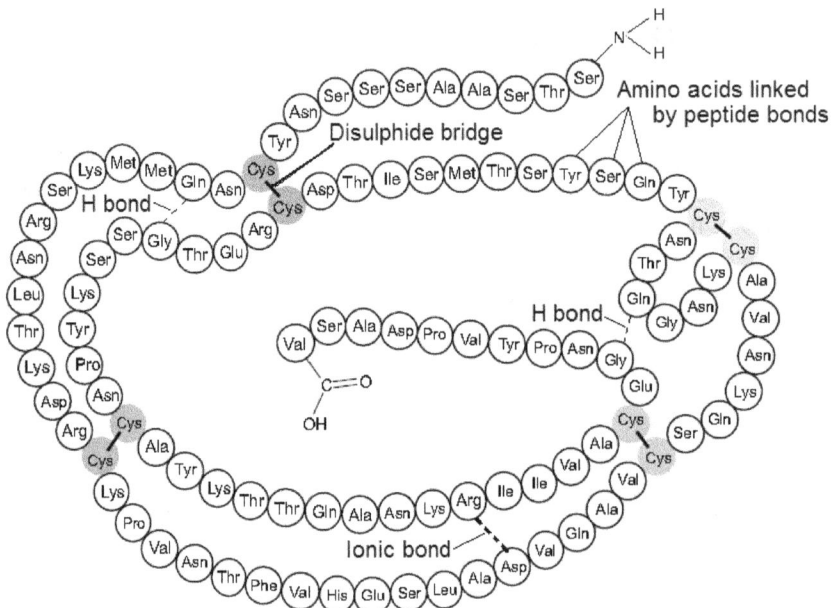

Fig.15.1

Enzyme specificity – Lock and key hypothesis

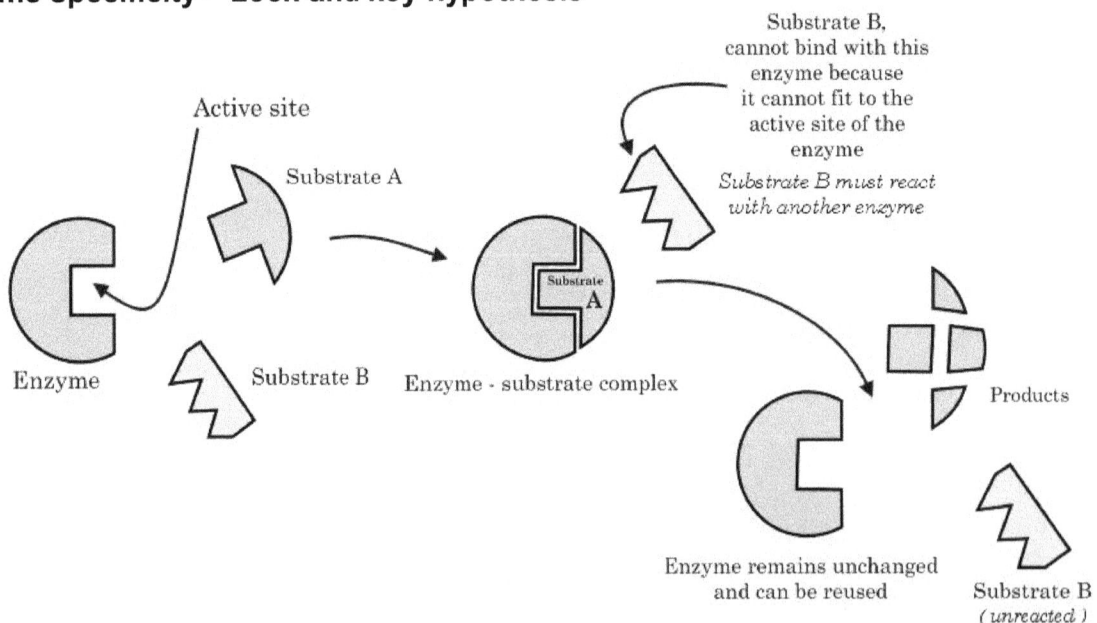

Fig.15.2

This specific tertiary shape is very essential for the functioning of enzymes. The part of the enzyme which reacts with the substrate is called the **active site**. The shape of the active site differs from one enzyme to another. This makes the enzyme react only with a **specific substrate**, which fits the active site, as in shown in fig 15.2.

Activation energy

Activation energy is the minimum energy that the reactant molecules must possess in order to start a reaction. Thus to start a reaction, energy must be supplied to the substrate molecules. This energy is called the activation energy. Enzymes lower the activation energy and provide an alternate, lower energy pathway for the reaction to proceed, as shown in fig.15.3. Thus the rate of reaction speeds up. So enzymes are referred to as **biological catalysts.**

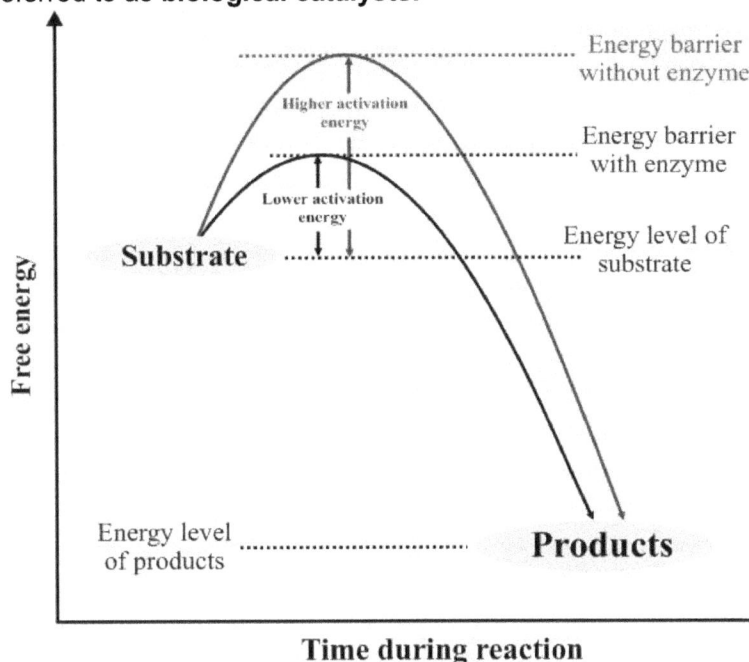

Fig.15.3

Induced fit hypothesis – lowering activation energy

The lock and key hypothesis explained the specificity of enzyme action, however it could not explain how enzymes lower the activation energy. The induced fit model of enzyme activity explains how enzymes reduce activation energy, as shown in fig.15.4.

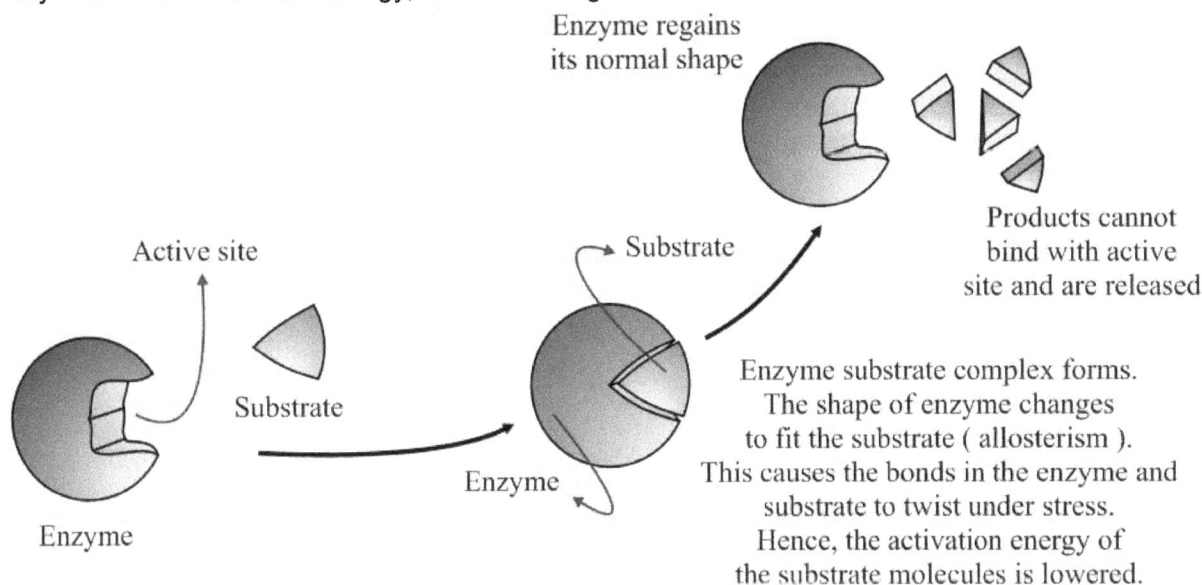

Fig.15.4

Other factors that could affect shape of enzymes
Temperature

Temperature can influence the shape of enzymes. As temperature increases up to the optimum, the rate of enzyme activity also increases. This is because enzyme and substrate molecules gain more kinetic energy and the collisions between active sites of enzymes and substrate molecule become *more frequent*. The *rate* of enzymes substrate complex formation increases so enzymes activity speeds up.

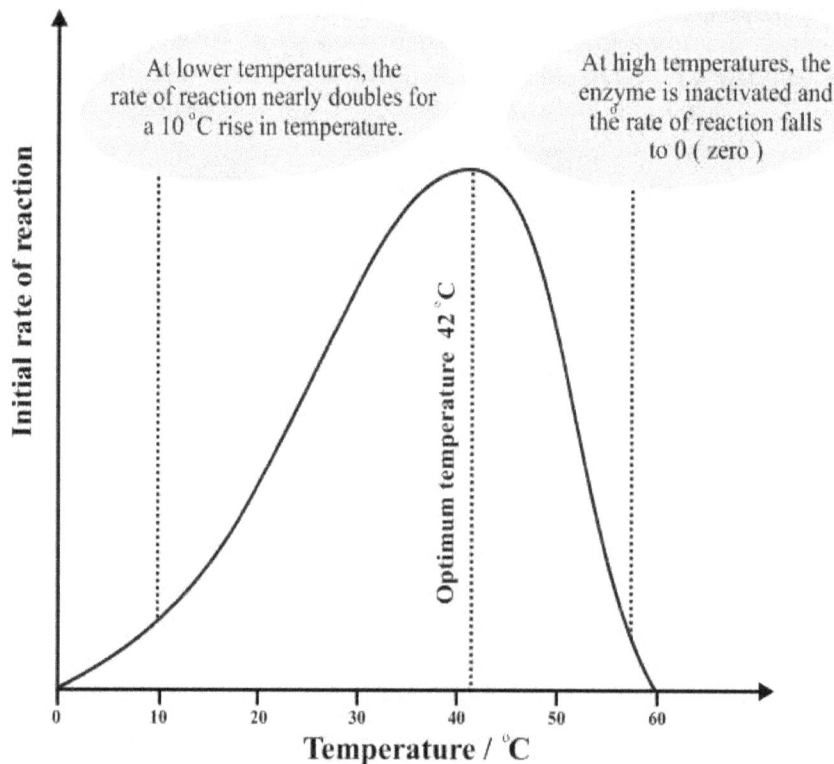

At lower temperatures, the rate of reaction nearly doubles for a $10\,°C$ rise in temperature.

At high temperatures, the enzyme is inactivated and the rate of reaction falls to 0 (zero)

Initial rate of reaction

Optimum temperature 42 °C

0   10   20   30   40   50   60

Temperature / °C

Fig.15.5

Beyond the optimum temperature, the rate of enzyme activity decreases because the high temperature causes the enzyme molecule to lose its specific 3D shape, due to breaking of Hydrogen bonds, ionic and covalent bonds. The active sites cannot bind with substrate, so enzyme substrate complexes cannot form. The optimum temperature of enzymes may differ depending upon the type of bonds that maintain the tertiary structure. Enzymes with many hydrogen bonds will denature easily as these bonds are relatively weak.

Non active site directed inhibitors

These are **non- competitive** inhibitors which bind to the enzyme molecule at an **allosteric site** and change the shape of the active site. This will prevent formation of enzyme substrate complex formation, reducing the rate of enzyme action.

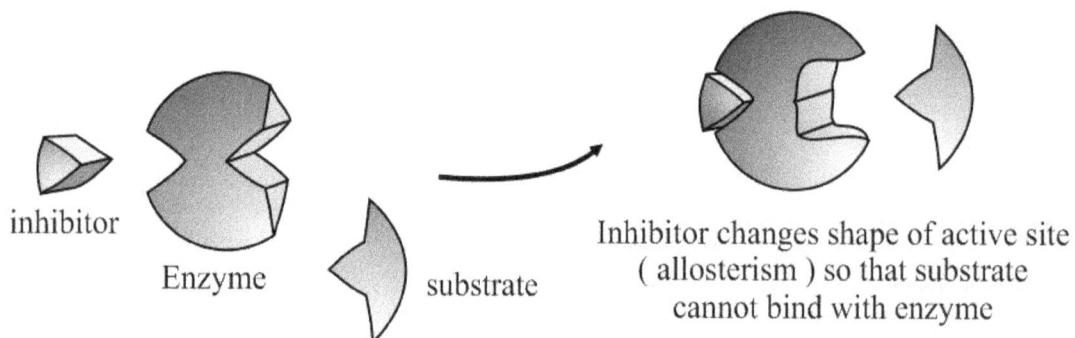

inhibitor

Enzyme

substrate

Inhibitor changes shape of active site (allosterism) so that substrate cannot bind with enzyme

Fig.15.6

Effect of pH on enzyme activity

Enzymes are very sensitive to changes in pH. Every enzyme works best within a very narrow range of pH. If the pH changes above or below the optimum then rate of reaction will decrease.

This is because at the optimum pH the H^+ ions / OH^- ionic balance (pH) of the solution is just right to maintain the specific 3D tertiary structure of the enzyme.

If pH changes then the change in H^+ ion concentration will disrupt the ionic bonds which maintain the tertiary structure of the enzyme. This will cause the enzyme to change its shape so that the active site cannot bind with substrate molecules. Enzyme substrate complexes will form at a slower rate, so rate of enzyme activity decreases.

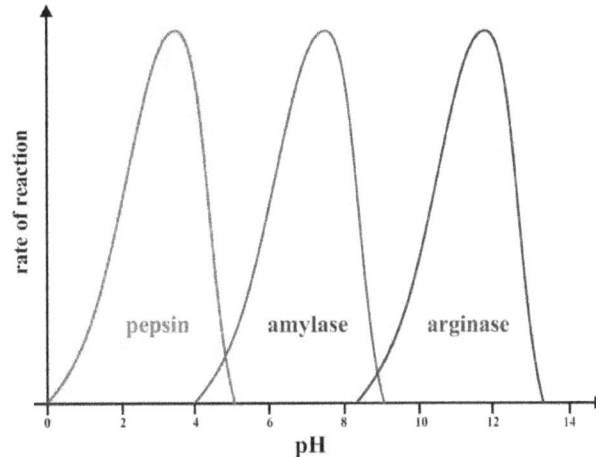

Fig.15.7

Roles of enzymes – controlling metabolic pathways

A metabolic pathway is a sequence of enzyme-controlled reactions, where every step in the pathway is controlled by a specific enzyme. These reactions may occur within cells (intracellular) or outside the cells (extracellular).

Digestion of starch is an example of extracellular metabolism.

Starch ──*Amylase(hydrolase)*──▸ **Maltose** ──*Maltase(hydrolase)*──▸ **Glucose**

Starch is the primary substrate, Maltose is an intermediate compound and glucose is the final product. Every reaction in the pathway is controlled by a specific enzyme. This enables the cell to have more control over metabolic reactions, by adjusting the concentration of enzymes to slow down or speed up specific reactions in the body. The reaction can be stopped by shutting down the secretion of the enzyme.

Glycolysis is an example of an intracellular metabolic pathway that occurs during respiration in the cytoplasm of cells. The *simplified* metabolic pathway below shows some of the enzymes involved in glycolysis.

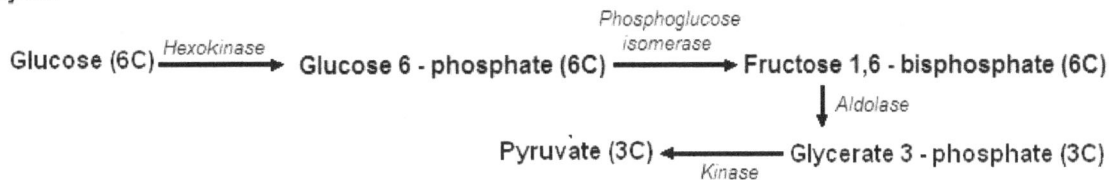

Glucose (6C) ──*Hexokinase*──▸ Glucose 6 - phosphate (6C) ──*Phosphoglucose isomerase*──▸ Fructose 1,6 - bisphosphate (6C)

│*Aldolase*↓

Pyruvate (3C) ◂────*Kinase*──── Glycerate 3 - phosphate (3C)

Classes of Enzymes

- o **Hydrolases** - general term for enzymes that catalyze a hydrolytic cleavage reaction.
- o **Isomerases** - catalyze the rearrangement of bonds within a single molecule.
- o **Kinases** - catalyze the addition of phosphate groups to molecules.
- o **Polymerases** - catalyze polymerization reactions such as the synthesis of DNA and RNA.

Note: The common name of an enzyme usually indicates the substrate and the nature of the reaction catalyzed. Enzyme names typically end in "-ase".

Core practical on enzyme concentration

Learning outcome: *Edexcel Syllabus Spec 9: Describe how enzyme concentrations can affect the rates of reactions and how this can be investigated practically by measuring the initial rate of reaction.*

AIM : To investigate how enzyme concentration can affect the rate of reaction.

PRINCIPLE: Protease enzymes will hydrolyse the solid proteins in Gelatin to soluble amino acids. Gelatin will hydrolyse rapidly as the concentration of enzyme increases.

MATERIALS REQUIRED: 75 g of 2% gelatin solution, pH 2.8 buffer, pepsin solution, funnel, 2 pipettes, stop clock, test tubes, test tube rack, spatula, Beaker, glassware for diluting enzymes, eye protection.

PROCEDURE:

1. Weigh 75 gm of solid freshly prepared gelatin jelly and place it into a beaker.
2. Pipette out 3 cm^3 of buffer solution to a test tube and add 7 cm^3 of 1% pepsin solution to this test tube and mix well.
3. Pour this solution to the beaker containing gelatin jelly. Stir the contents carefully for 15 sec. using a spatula.
4. Leave the mixture to stand for 5 min at a temperature of 35^0C.
5. Suspend a funnel over an empty conical flask.
6. Pour the mixture in to the funnel, using the spatula to empty the beaker.
7. Start the stopwatch as the first drop leaves the funnel.
8. Record the time taken for the jelly to pass through the funnel completely.
9. Repeat the experiment with a range of different enzyme concentrations.
10. The control setup will use distilled water instead of enzyme solution.
11. Tabulate the results and plot graph.

Enzyme concentration: As the enzyme concentration increases (at constant substrate concentration) the rate of reaction increases until it reaches a maximum rate (V max). This is because there will be more number of free active sites, at any given time. So the rate of enzyme substrate complex formation increases. Thus rate of reaction increases. The rate doesn't increase beyond the V max because the substrate concentration becomes a limiting factor. Even though there will be many free active sites, there will not be enough substrate molecules to bind with them. So, rate of enzyme substrate complex formation remains constant at V max. But, increasing substrate concentration would further increase the rate of reaction.

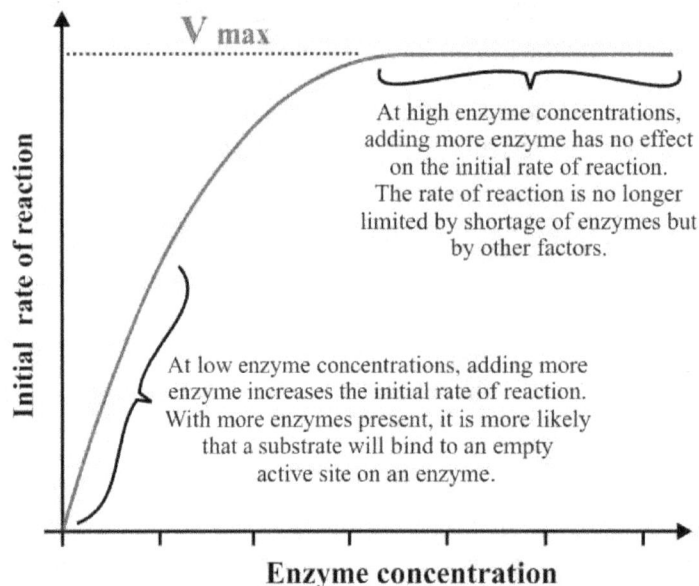

Fig.15.8

Note: If a fixed quantity of enzyme and substrate are allowed to react for a period of time, the rate of enzyme action will decrease because the substrate gets used up and the substrate concentration decreases over time. So, to get an accurate estimation of the rate of enzyme action, the initial rate can be measured with excess substrate for a short duration of time, before the substrate concentration becomes a limiting factor.

CHAPTER SIXTEEN
NUCLEIC ACIDS

Learning outcomes: by the end of this chapter you should be able to
- *Edexcel Syllabus Spec 10: Describe the basic structure of mononucleotides (as a deoxyribose or ribose linked to a phosphate and a base, i.e. thymine, uracil, cytosine, adenine or guanine) and the structures of DNA and RNA (as polynucleotides composed of mononucleotides linked through condensation reactions) and describe how complementary base pairing and the hydrogen bonding between two complementary strands are involved in the formation of the DNA double helix.*

Mononucleotide structure and formation

DNA and RNA belong to a group of compounds called **Nucleic acids**. The building blocks of nucleic acids are called **mononucleotides.** Ribonucleic acid (RNA) and deoxyribonucleic acid (DNA) are made up by linking of many mononucleotides, by condensation reactions.

A mononucleotides is made up of three parts, as shown in fig.16.1.
- An inorganic phosphate group;
- A pentose sugar - ribose in RNA and deoxyribose in DNA;
- A base which contains nitrogen - Adenine (A), Guanine (G), Cytosine (C), Uracil (U) **(only in RNA)**, Thymine (T) **(only in DNA)**

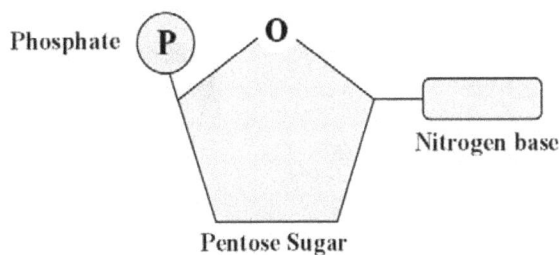

Fig.16.1

Condensation reactions join up the inorganic phosphate group and nitrogenous base to the sugar by removal of water molecules, as shown in **fig 16.2**.

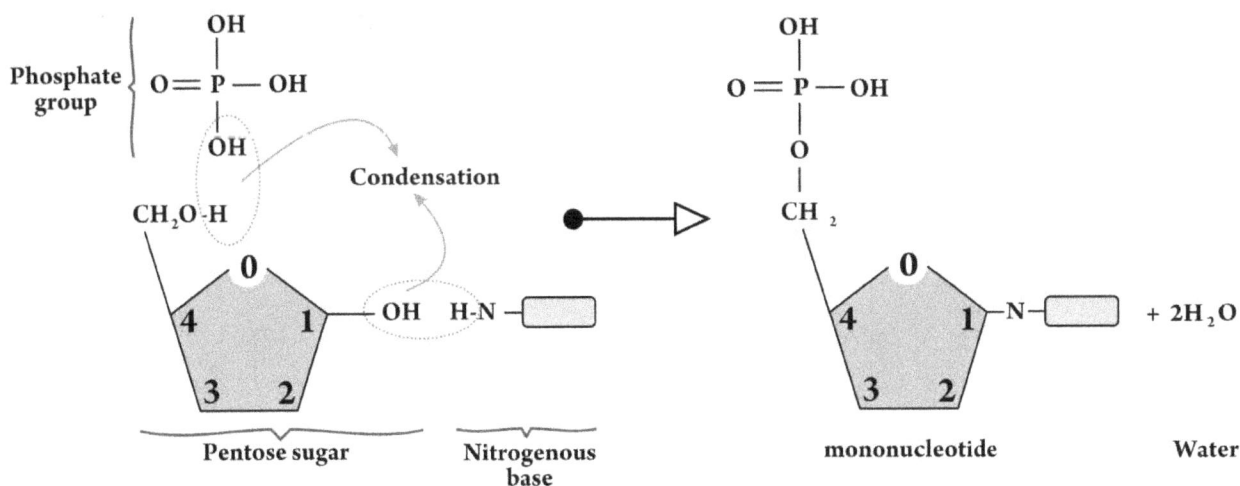

Fig.16.2

Formation of a polynucleotide by condensation

Condensation reactions are involved in the formation of phospho-diester bonds. Carbon 5' of the pentose binds with carbon 3' of pentose on another nucleotide. A phosphate group links the two pentose molecules to each other. Many nucleotides can be linked by phospho-diester bonds to form a polynucleotide.

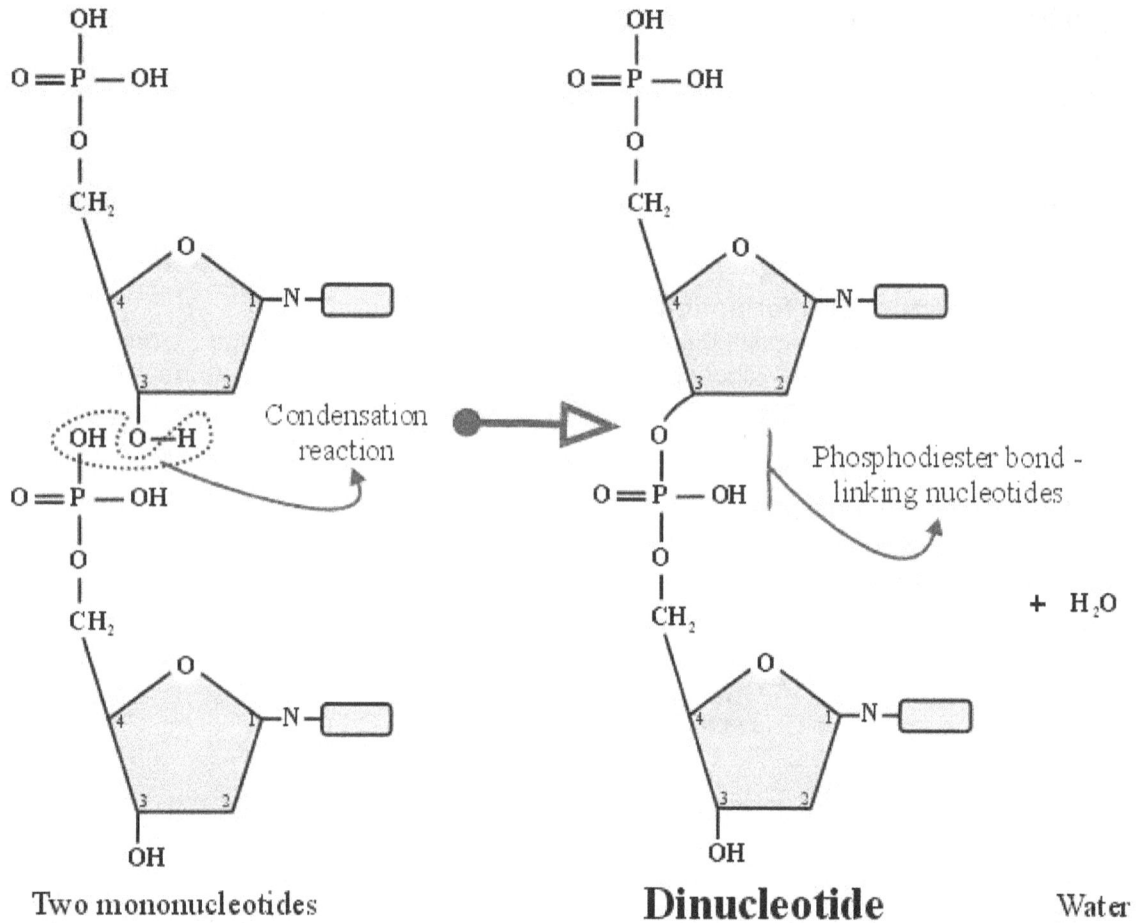

Two mononucleotides   Dinucleotide   Water

Fig.16.3

Structure of RNA

RNA is a single polynucleotide chain. The mononucleotides are linked to each other by phospho-diester bonds. Each mononucleotide contains

- **a ribose sugar,**
- **a nitrogenous base (except thymine)** and
- **a phosphate group.**

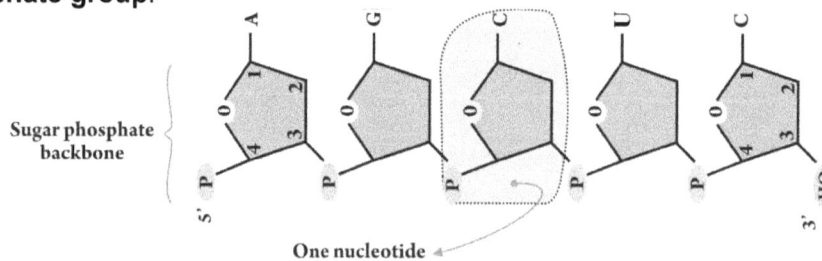

Fig.16.4

Structure of DNA

DNA is a polymer of deoxyribo-nucleotides. Each mononucleotide in DNA consists of
- **a deoxyribose sugar**,
- **a nitrogenous base (except uracil)** and
- **a phosphate group**.

DNA consists of two polynucleotide strands which are anti parallel, one strand runs from 3' to 5' end and the other runs from 5' to 3' end. The two strands are linked to each other by **Hydrogen bonds between nitrogenous base pairs**, Adenine pairs with Thymine and Guanine pairs with cytosine. The strands are wound in to a **double helix**. The base pairs are 0.34 nm apart and there are 10 base pairs in one complete turn of the helix. Both strands are complementary to each other.

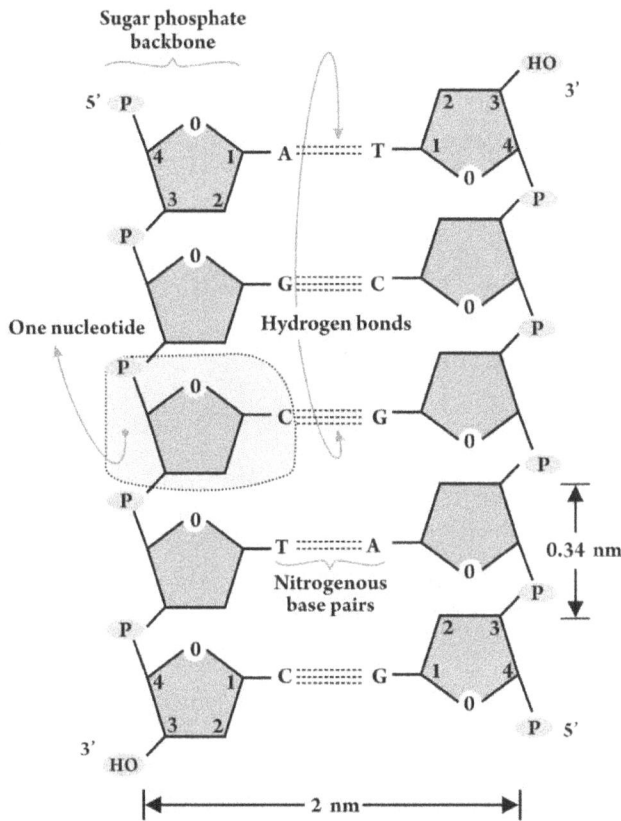

Fig.16.5

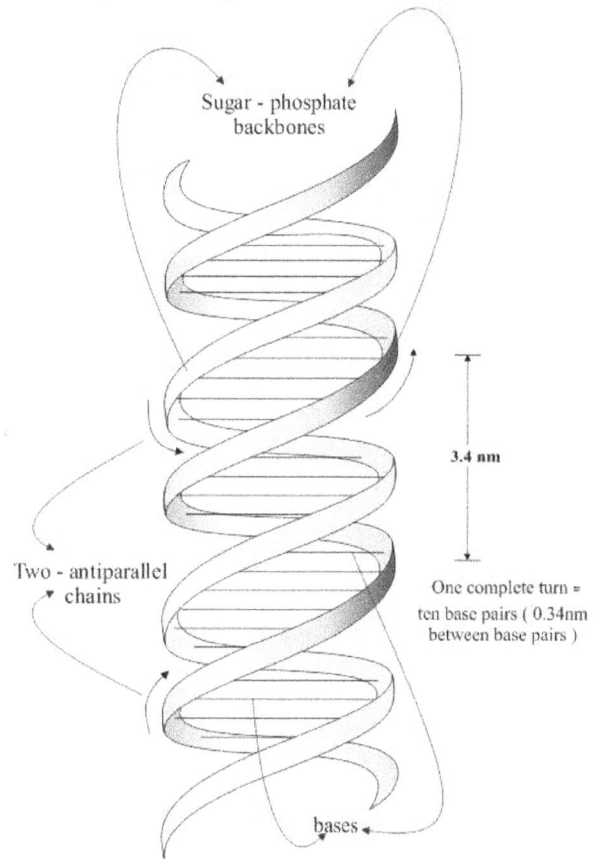

Fig.16.6

Learning outcomes: by the end of this chapter you should be able to
- **Edexcel Syllabus Spec 11:** *Explain the nature of the genetic code (triplet code only; nonoverlapping and degenerate not required at AS).*

DNA Replication

During semi conservative replication of DNA, each daughter DNA molecule consists of one complete old strand and one complete new strand of nucleotides, as shown in fig.17.1.

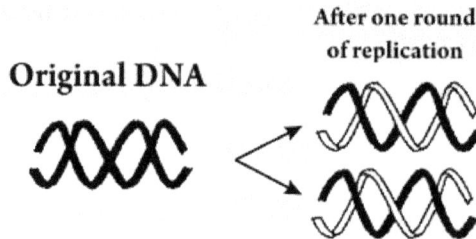

Original DNA

After one round of replication

Fig.17.1

The process of semi conservative replication of DNA is illustrated in fig.17.2.

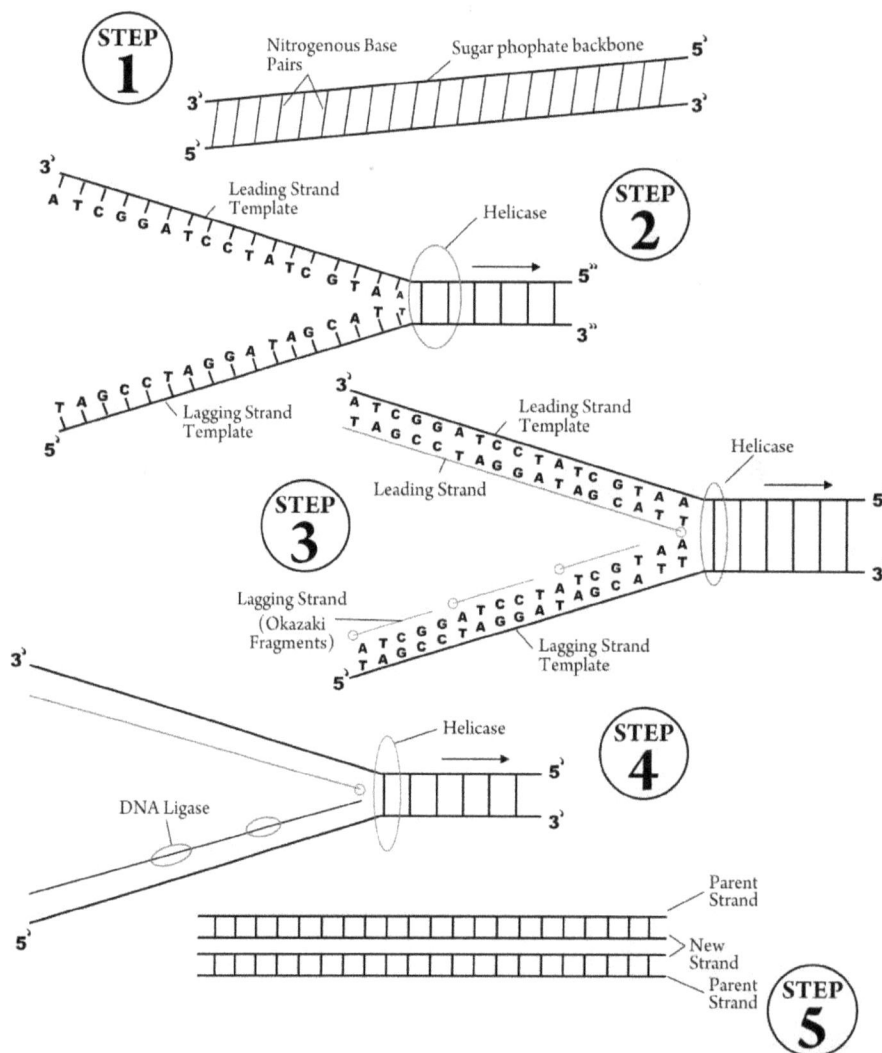

STEP 1

Nitrogenous Base Pairs

Sugar phophate backbone

STEP 2

Leading Strand Template

Helicase

Lagging Strand Template

STEP 3

Leading Strand Template

Leading Strand

Helicase

Lagging Strand (Okazaki Fragments)

Lagging Strand Template

STEP 4

Helicase

DNA Ligase

Parent Strand

New Strand

Parent Strand

STEP 5

Fig.17.2

1. The two strands of DNA are held together by hydrogen bonds between the nitrogenous base pairs.
2. The enzyme *Helicase* unwinds and unzips the DNA by breaking the hydrogen bonds between the nitrogenous base pairs. This exposes the nitrogenous bases on each strand.
3. Each DNA strand will serve as a **template** for free nucleotides to pair up with complementary nitrogenous bases, following the base pairing rule – Adenine with Thymine and Guanine with Cytosine. The enzyme *DNA polymerase* can then join the nucleotides to each other by **phospho-diester bonds.**
4. On the leading strand, the nucleotides are joined continuously but on the lagging strand, the new polynucleotide chain is formed in fragments *(Okazaki Fragments)*. These fragments are joined up by *DNA ligase.*
5. Two identical molecules of daughter DNA are produced, but each daughter DNA molecule has one complete parent strand and one complete new strand of polynucleotides. This is called **semi-conservative replication.**

Competing theories
The mechanism of DNA replication was postulated in the form of three theories.
1. Conservative replication: This theory suggested that a "photocopy" of the original DNA was made, leaving the original DNA conserved.
2. Semi-conservative replication: This theory suggested that each new DNA molecule contains one complete new strand and one complete old strand.
3. Dispersive replication: This theory suggested that each daughter DNA molecule consisted of fragments of old and new nucleotide strands.

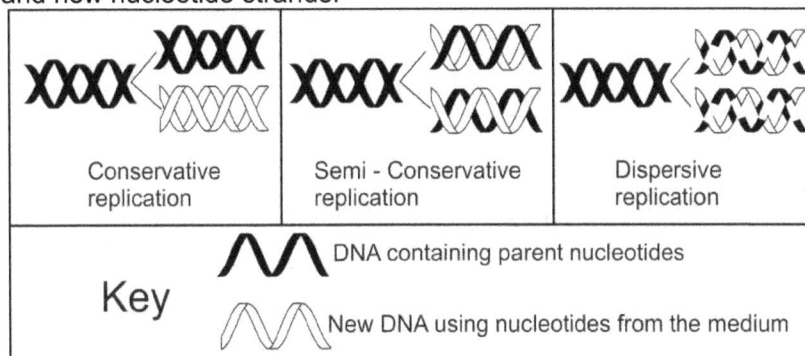

Fig.17.3

The proof that the semi-conservative method was the correct method came from an experiment performed by Meselson and Stahl using the bacterium *E. coli* together with the technique of density gradient centrifugation, which separates molecules on the basis of their density.

Meselson and stahl density gradient experiment
Density gradient in caesium chloride solution
When a caesium chloride soultion is centrifuged, a density gradient is setup. The upper layers in the centrifuge tube have a minimum density. The density of the solution increases as we go deeper into the solution. DNA is insoluble in caesium chloride and will settle into the soultion at the depth where the density od DNA matches the density of the solution. So, denser DNA will settle down deeper in the solution and lighter, less dense DNA will settle down just below the surface, where the density of the solution is lower.

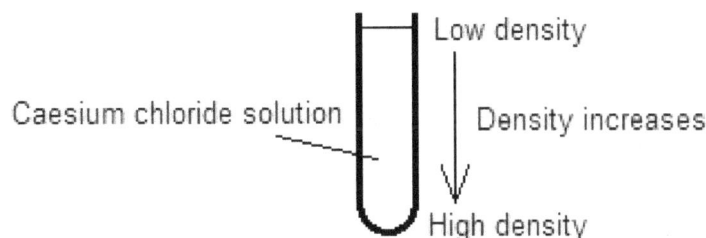

Fig.17.4

The experimental techniques and results

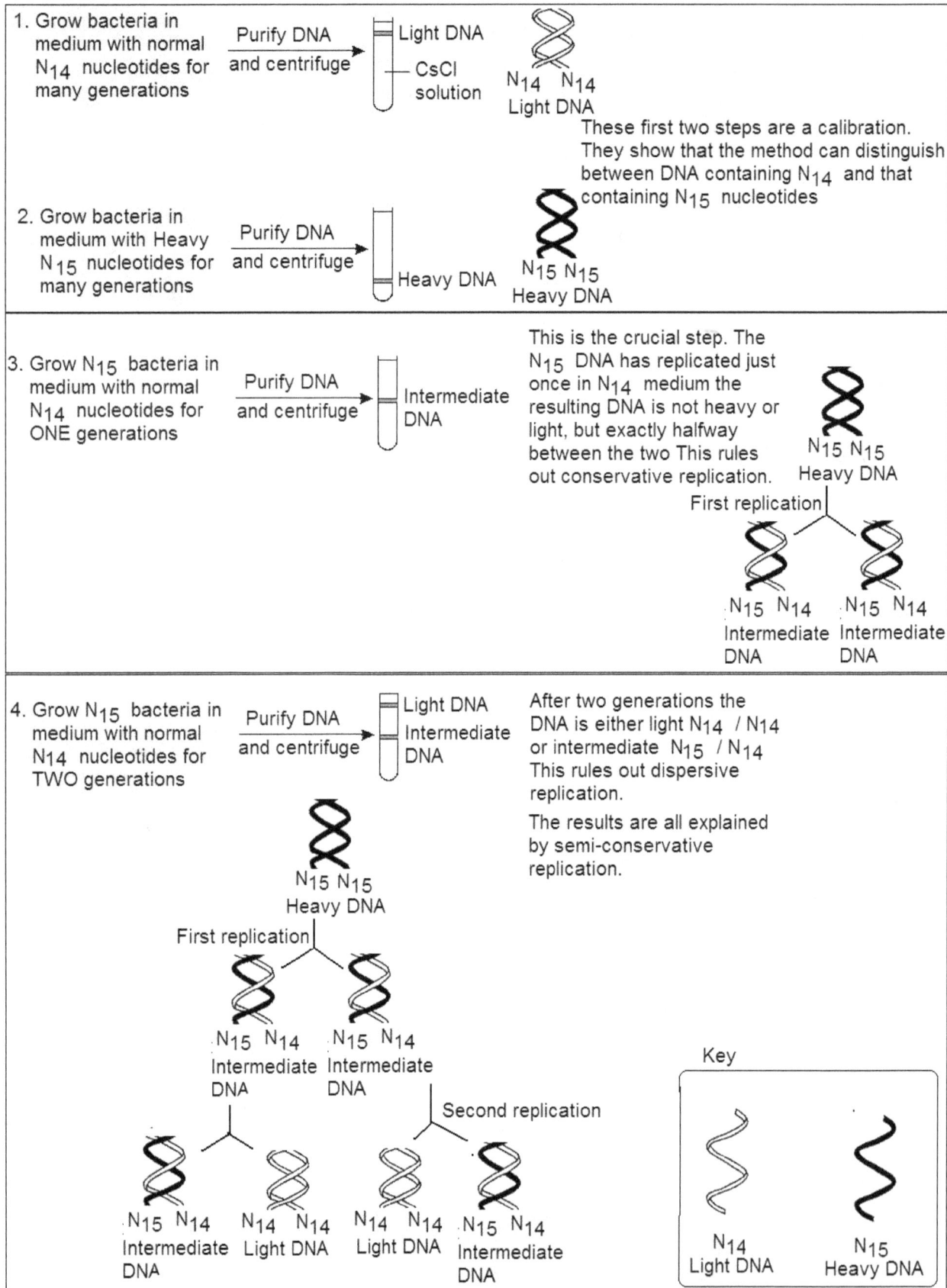

1. Grow bacteria in medium with normal N_{14} nucleotides for many generations → Purify DNA and centrifuge → Light DNA — CsCl solution — N_{14} N_{14} Light DNA

These first two steps are a calibration. They show that the method can distinguish between DNA containing N_{14} and that containing N_{15} nucleotides

2. Grow bacteria in medium with Heavy N_{15} nucleotides for many generations → Purify DNA and centrifuge → Heavy DNA — N_{15} N_{15} Heavy DNA

3. Grow N_{15} bacteria in medium with normal N_{14} nucleotides for ONE generations → Purify DNA and centrifuge → Intermediate DNA

This is the crucial step. The N_{15} DNA has replicated just once in N_{14} medium the resulting DNA is not heavy or light, but exactly halfway between the two This rules out conservative replication.

N_{15} N_{15} Heavy DNA

First replication

N_{15} N_{14} Intermediate DNA N_{15} N_{14} Intermediate DNA

4. Grow N_{15} bacteria in medium with normal N_{14} nucleotides for TWO generations → Purify DNA and centrifuge → Light DNA / Intermediate DNA

After two generations the DNA is either light N_{14} / N_{14} or intermediate N_{15} / N_{14} This rules out dispersive replication.

The results are all explained by semi-conservative replication.

N_{15} N_{15} Heavy DNA

First replication

N_{15} N_{14} Intermediate DNA N_{15} N_{14} Intermediate DNA

Second replication

N_{15} N_{14} Intermediate DNA N_{14} N_{14} Light DNA N_{14} N_{14} Light DNA N_{15} N_{14} Intermediate DNA

Key

N_{14} Light DNA N_{15} Heavy DNA

Fig.17.5

Chapter Eighteen
The Genetic code and Protein synthesis

Learning outcomes: by the end of this chapter you should be able to

- ***Edexcel Syllabus Spec 12:*** *Explain the nature of the genetic code (triplet code only; nonoverlapping and degenerate not required at AS).*
- ***Edexcel Syllabus Spec 13:*** *Describe a gene as being a sequence of bases on a DNA molecule coding for a sequence of amino acids in a polypeptide chain.*
- ***Edexcel Syllabus Spec 14:*** *Outline the process of protein synthesis, including the role of transcription, translation, messenger RNA, transfer RNA and the template (antisense) DNA strand (details of the mechanism of protein synthesis on ribosomes are not required at AS).*

The Genetic Code – the sequence of bases on Nucleic acids that code for amino acids.
The genetic information on the Nucleic acids (DNA and RNA) is in the form of nitrogenous bases. The sequence of bases on DNA determines the sequence of amino acids in a polypeptide chain (primary protein structure). **The sequence of nitrogenous bases which code for amino acids is called as the genetic code**. The fig.18.1shows the codons on **mRNA** that code for specific amino acids.

Second Letter

First Letter	U	C	A	G	Third Letter
U	UUU UUC Phenyl-alanine UUA UUG Leucine	UCU UCC UCA UCG Serine	UAU UAC Tyrosine UAA Stop codon UAG Stop codon	UGU UGC Cysteine UGA Stop codon UGG Tryptophan	U C A G
C	CUU CUC CUA CUG Leucine	CCU CCC CCA CCG Proline	CAU CAC Histidine CAA CAG Glutamine	CGU CGC CGA CGG Arginine	U C A G
A	AUU AUC AUA Isoleucine AUG Methionine; start codon	ACU ACC ACA ACG Threonine	AAU AAC Asparagine AAA AAG Lysine	AGU AGC Serine AGA AGG Arginine	U C A G
G	GUU GUC GUA GUG Valine	GCU GCC GCA GCG Alanine	GAU GAC Aspartic acid GAA GAG Glumatic acid	GGU GGC GGA GGG Glycine	U C A G

Fig.18.1

Triplet code
If the code is made up of one base it could code for 4 amino acids, as shown in fig.18.2.

Genetic code (Single base)	Amino acid
A	XXX
T	XXX
C	XXX
G	XXX

Fig.18.2

If the code is made up of two bases it could code for 16 amino acids, as shown in fig.18.3.

Genetic code (Double base)	Amino acid	Genetic code (Double base)	Amino acid	Genetic code (Double base)	Amino acid	Genetic code (Double base)	Amino acid
AA	XXX	TA	XXX	CA	XXX	GA	XXX
AT	XXX	TT	XXX	CT	XXX	GT	XXX
AC	XXX	TC	XXX	CC	XXX	GC	XXX
AG	XXX	TG	XXX	CG	XXX	GG	XXX

Fig.18.3

Twenty different amino acids are needed to make all the proteins that are needed by living organisms. A single base code or a double base code would not be sufficient enough to code for all the twenty amino acids. A **triplet base code would be able to code for all the 20 amino acids**. So, the bases are read in groups of **3 (Triplet code)**. This gives 64 combinations, more than enough to code for 20 amino acids. The genetic code on the antisense strand of DNA is shown in fig.18.4.

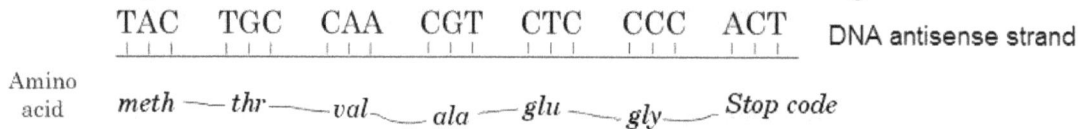

TAC   TGC   CAA   CGT   CTC   CCC   ACT   DNA antisense strand

Amino acid   *meth* —— *thr* —— *val* —— *ala* — *glu* —— *gly* —— *Stop code*

Fig.18.4

Deciphering the code

In the early 1960's Nirenberg devised a method of producing polypeptides by using synthetic mRNA in a cell-free system. The procedure is illustrated in fig.18.5.

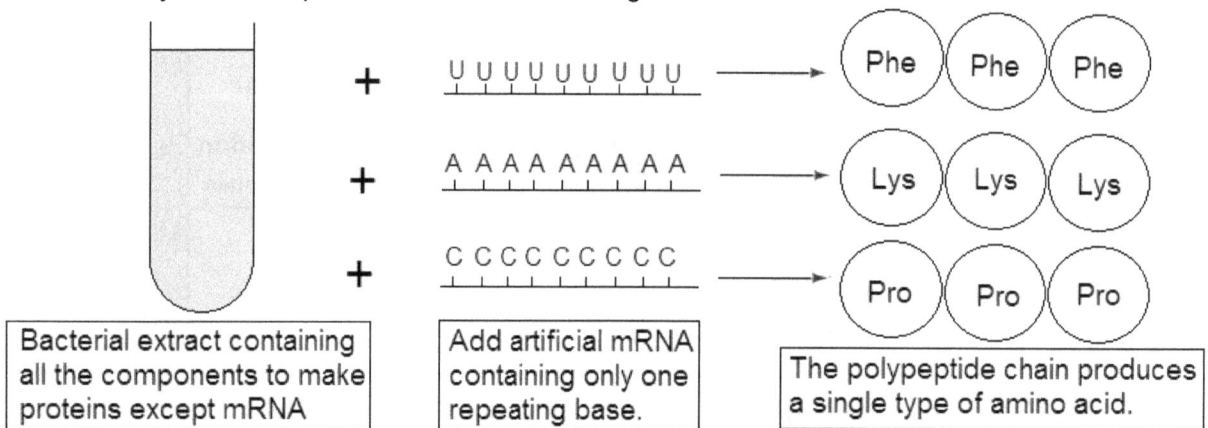

| Bacterial extract containing all the components to make proteins except mRNA | Add artificial mRNA containing only one repeating base. | The polypeptide chain produces a single type of amino acid. |

Fig.18.5

The results of this experiment paved the way for determining the genetic code and also helped to prove that the genetic code is **a triplet code**, as the number of the **bases on mRNA was three times greater than the number of amino acids in the polypeptide chain**.

Gene

A gene is **a sequence of bases on a DNA molecule coding for a sequence of amino acids in a polypeptide chain**. Genes express themselves by specifying the sequence of amino acids in polypeptide chains during **protein synthesis,** as shown in fig. 18.6.

The genes control metabolic activity by producing specific **enzymes** needed for metabolism. It may also produce specific **functional proteins** like Insulin, Haemoglobin, Myoglobin, Rhodopsin and Glucagon. Genes also specify the amino acid sequence of **structural proteins** like keratin (in nails), collagen (in bones), actin and myosin (in muscles), carrier and channel proteins.

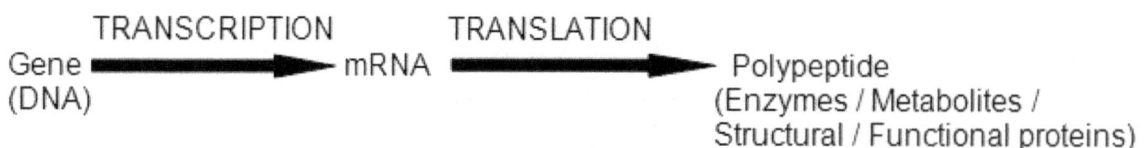

TRANSCRIPTION   TRANSLATION
Gene ➡️ mRNA ➡️ Polypeptide
(DNA)   (Enzymes / Metabolites /
                                                              Structural / Functional proteins)

Fig. 18.6

Protein synthesis

Protein synthesis is the formation of a polypeptide chain by using the information on the DNA. The information from the gene (DNA) is copied on to mRNA in the nucleus by a process called **transcription**. The mRNA is then used on the ribosomes to form a polypeptide chain by a process called **translation**.

Transcription - making **mRNA from DNA**

1. The enzyme RNA polymerase attaches to the promoter region of DNA and begins to unwind the **gene** (cistron) by breaking the hydrogen bonds between the base pairs.
2. Free ribonucleotides in the nucleus pair up with complementary bases on the **antisense strand** of DNA. **A** pairs with **U**; **G** pairs with **C**; **T** pairs with **A**.
3. The free nucleotides then get linked to each other by the formation of phospho-diester bonds, formed by condensation reactions. It is catalysed by RNA polymerase.
4. The newly formed strand of mRNA then peels away from the coding strand and the DNA rewinds. The process continues until a stop codon is reached.
5. A primary transcript of mRNA or pre-mRNA is formed. It contains introns and exons.
6. Before leaving the nucleus, introns of mRNA are cut off. The remaining nucleotides rejoin and are called exons. The exons (mature mRNA) leave the nucleus.
7. Poly adenine tail and guanine cap are attached to the mRNA before it diffuses out of the nucleus through the nuclear pores. A length of DNA, called as a gene or cistron, is copied onto a single stranded mRNA molecule during transcription.

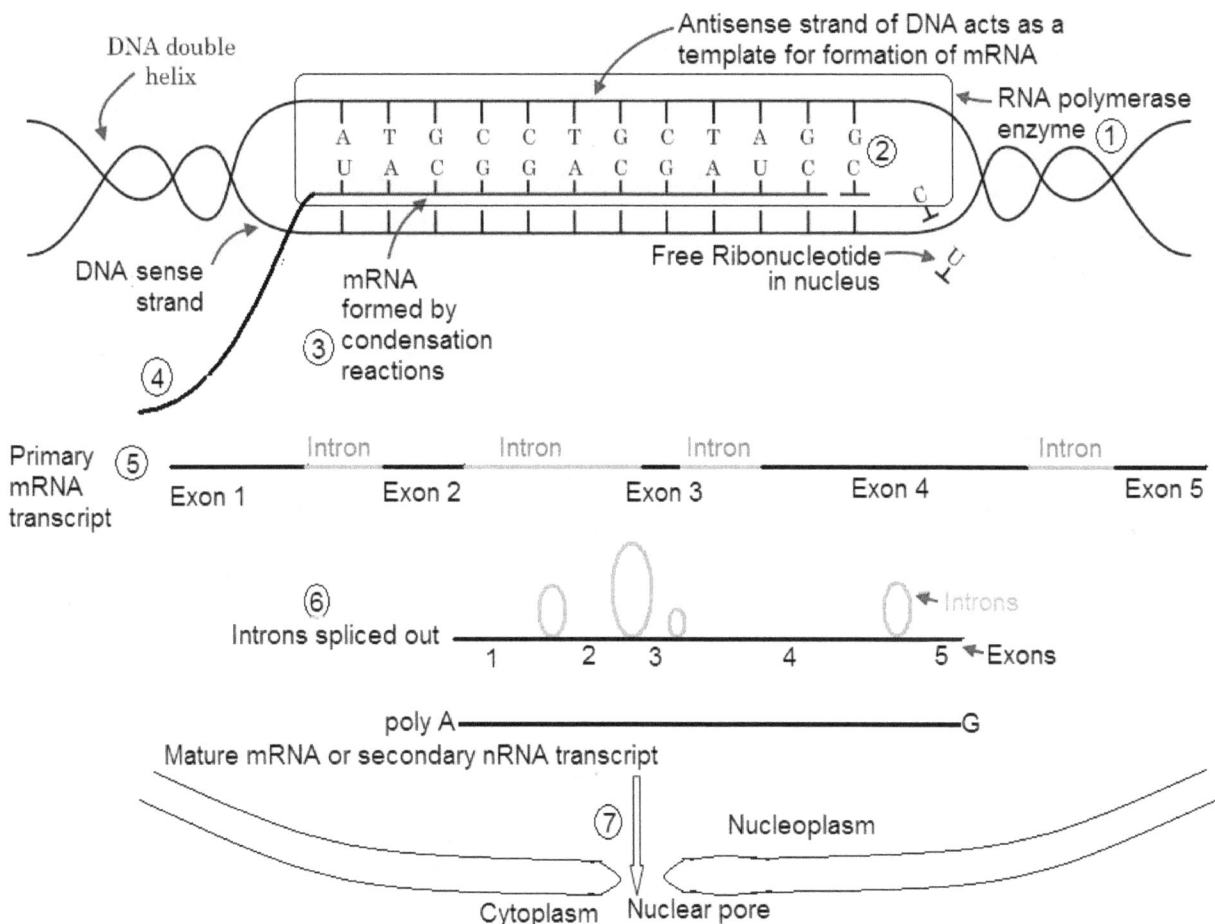

Fig.18.7

Translation – decoding genetic information to form a polypeptide

Using the genetic information (sequence of bases) on mRNA to form a polypeptide chain with a specific sequence of amino acids is called translation. It occurs on the ribosomes in the cytoplasm.

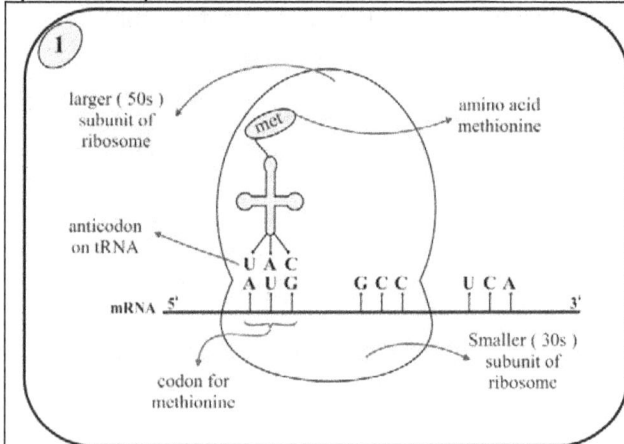

Fig.18.8

A ribosome attaches to the mRNA at an initiation codon (AUG). The ribosome encloses two codons. met-tRNA diffuses to the ribosome and the anticodon attaches to the mRNA start codon by complementary base pairing.

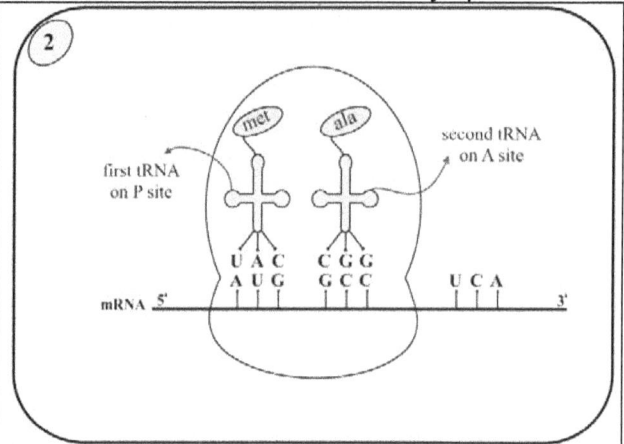

Fig.18.9

The next amino acid-tRNA attaches to the adjacent mRNA codon (ala in this case).

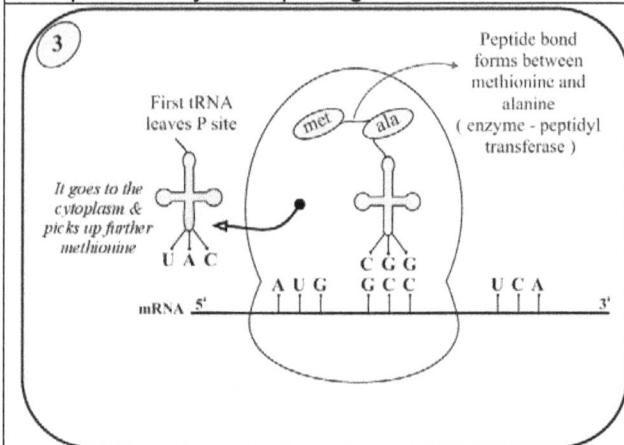

Fig.18.10

The bond between the amino acid and the tRNA is cut and a peptide bond is formed between the two amino acids. The free tRNA molecule leaves to collect another amino acid.

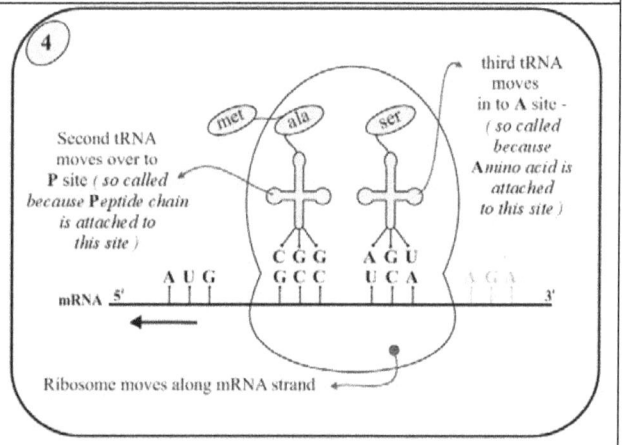

Fig.18.11

The ribosome moves along one codon so that a new amino acid-tRNA can attach. The polypeptide chain elongates one amino acid at a time, and peels away from the ribosome, folding up into a protein as it goes. This continues until a stop codon is reached.

Note: *The m*ain function of the ribosome is to hold the mRNA molecule so that anticodons of tRNA can pair up with complementary codons of mRNA. This brings amino acids to lie adjacent to each other so that peptide bonds can be formed.

A single piece of mRNA can be translated by many ribosomes simultaneously. A group of ribosomes all attached to one piece of mRNA is called a polysome, as shown in fig.18.12.

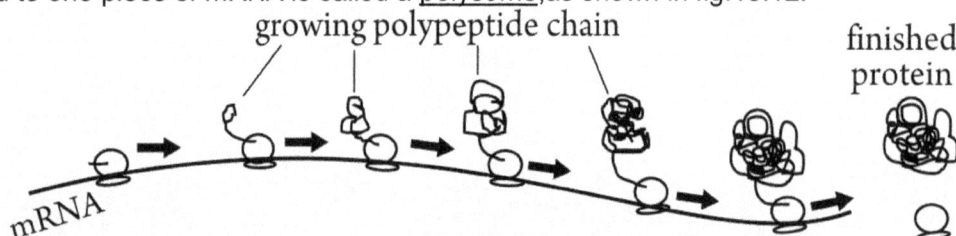

Fig.18.12

Structure and role of mRNA *(messenger RNA)*

mRNA is a single chain of polynucleotides. The nucleotides are linked to each other by phospho-diester bonds. The number of nucleotides is thrice as much as the amino acids in the polypeptide chain that it codes for.

The nucleotides contain
- **ribose sugar**
- phosphate and
- the nitrogenous base A, **U**, G or C [**no Thymine (T) is present**].

Function

mRNA is a complementary copy of the antisense strand of DNA. Its function is to **copy and carry** genetic information from the DNA inside the nucleus to the ribosomes in the cytoplasm for proteins synthesis (translation). It is formed during transcription by copying genetic information from the **antisense DNA strand**.

The genetic code can also be read from mRNA strand. A sequence of three nitrogenous bases on mRNA which codes for a specific amino acid is called a **CODON**.

Fig.2.58 shows the relationship between the **triplet bases** on DNA, the **codons** on mRNA and the amino acids that are coded for by the respective bases.

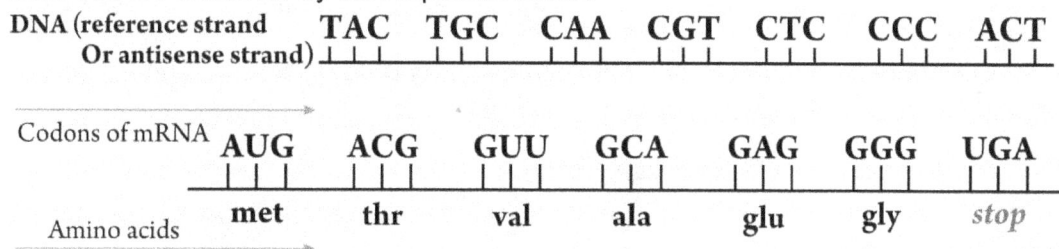

DNA (reference strand Or antisense strand)	TAC	TGC	CAA	CGT	CTC	CCC	ACT
Codons of mRNA	AUG	ACG	GUU	GCA	GAG	GGG	UGA
Amino acids	met	thr	val	ala	glu	gly	*stop*

Fig.18.13

Structure and role of tRNA (Transfer RNA)

tRNA is a single chain of polynucleotides. The nucleotides are linked to each other by phospho-diester bonds. The nucleotides have the same composition as mRNA nucleotides.

tRNA is only about 80 nucleotides long, and it folds up by complementary base pairing to form a clover-leaf structure. At one end of the molecule there is an **amino acid binding site**. On the middle loop there is a triplet nucleotide sequence called the **anticodon**, as shown in fig.18.14. There are 61 different tRNA molecules, each with a different anticodon sequence complementary to the 61 different codons on mRNA. The remaining three codons are stop codons.

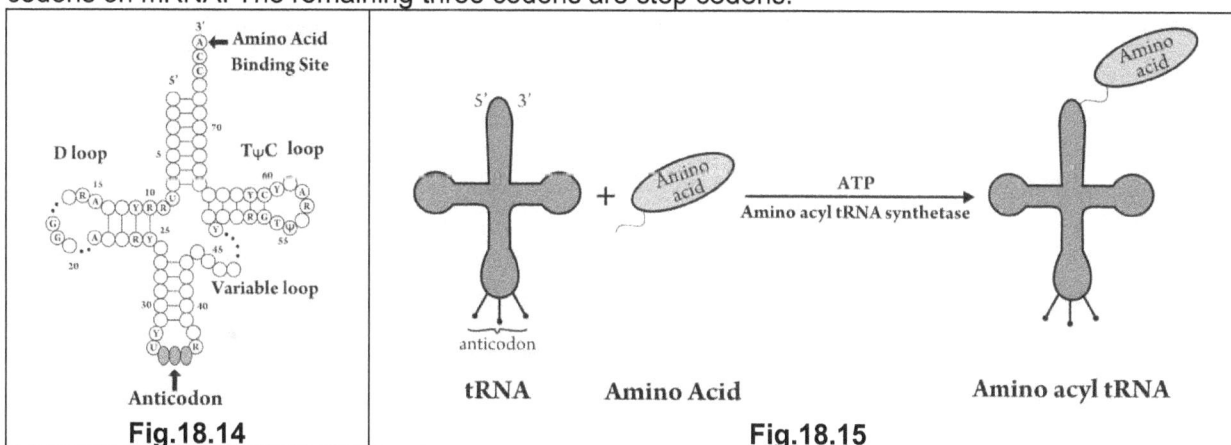

Fig.18.14 **Fig.18.15**

The main function of tRNA molecule is to transfer amino acids from the cytoplasm to the ribosome. The tRNA then matches amino acids to their specific codons within the ribosome. The tRNA **binds with a specific amino acid in the cytoplasm** to form an amino-acyl tRNA complex catalysed by the enzyme amino-acyl tRNA synthetase as shown in fig.18.15. The amino acid to which tRNA binds is determined by the anticodon of the tRNA. The activated **amino acid is then transferred to the ribosome**. The anticodons will pair up with complementary codons on mRNA. This helps to build up a specific sequence of amino acids in the polypeptide chain.

CHAPTER NINETEEN
MUTATIONS

Learning outcomes: by the end of this chapter you should be able to
- **Edexcel Syllabus Spec 15:** *Explain how errors in DNA replication can give rise to mutations and explain how cystic fibrosis results from one of a number of possible gene*

Mutation

A mutation is a **heritable change in the DNA of a cell**. Gene or **point mutations** result from a **change in the base sequence within a gene, usually during DNA replication**.

If mutations occur in **gamete cells**, then they can be passed on to successive generations. This is called as **germ line mutations**. Gene mutations may result in genetic disorders like, cystic fibrosis, Albinism and thalasaemia.

Mutations that occur in body cells cannot be passed onto successive generations. These mutations can only be passed onto daughter cells formed by mitosis from the mutant cell. These are called as **somatic mutations.** Eg: cancer cells.

Mutations occur **randomly and spontaneously** in nature, during DNA replication. The frequency with which one allele mutates to another is known as the **mutation rate**.

The flowchart below illustrates the effects of gene mutations.

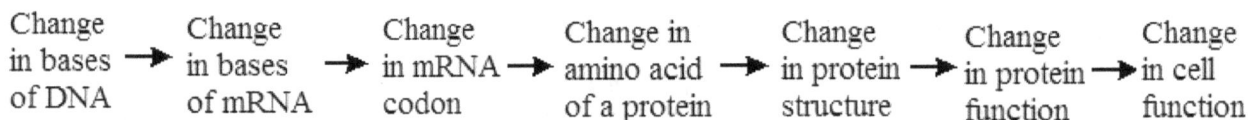

Change in bases of DNA → Change in bases of mRNA → Change in mRNA codon → Change in amino acid of a protein → Change in protein structure → Change in protein function → Change in cell function

A few types of gene mutations are described below

1. Neutral Mutations: occur in two ways.

a.) A codon is changed, but the **new codon still codes for the same amino acid** e.g. AUU, AUC and AUA all code for the amino acid isoleucine. Changing the last letter makes **no difference in the polypeptide chain**.

b.) A codon is changed resulting in a new amino acid in the protein. However, the different amino acid **does not change the shape of the protein**, and the activity of the protein is unaffected.

2. Inversion

In an inversion mutation, an entire section of DNA is reversed. A small inversion may involve only a few bases within a gene, while longer inversions involve large regions of a chromosome containing several genes. The inverted segment has been underlined in fig.19.1.

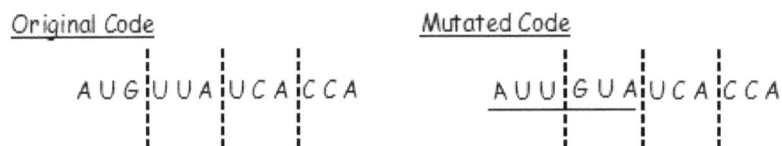

Original Code   Mutated Code

A U G | U U A | U C A | C C A   A U U | G U A | U C A | C C A

Fig.19.1

3. Substitution mutations

Substitution mutations occur where **one** letter is substituted for another. These mutations can often be neutral. A is substituted by U in the sixth base, as shown in fig.19.2.

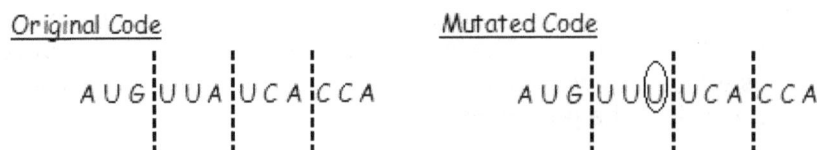

Original Code   Mutated Code

A U G | U U A | U C A | C C A   A U G | U U U | U C A | C C A

Fig.19.2

4. Insertion and Deletion mutations

These mutations occur when one (or more) letters are **added** or **deleted** from the DNA code. Inserting C between G and U has caused the bases to shift to the right. This is called **frame shift**. All the codons beyond the insertion will change. The original codons have been underlined in fig.19.3.

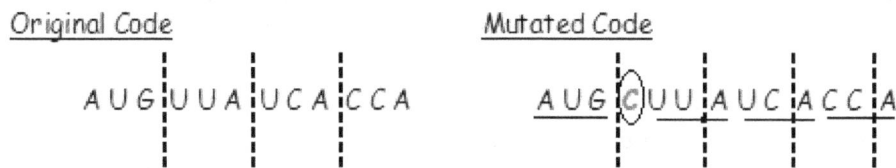

Original Code   Mutated Code

A U G ¦U U A ¦U C A ¦C C A   A U G ⓒU U ¦A U C ¦A C C ¦A

Fig.19.3

Frame-shift mutations occur after the addition or deletion of a base in the code. After the change all subsequent codons are changed, meaning the protein is usually severely affected.
Substitution may result in one or two amino acids being altered in the polypeptide chain. Substitution does not result in **frame shifts**.

5. Non-sense mutations

These mutations occur when a stop codon is accidentally created in the middle of a gene. This will result in only half of the code being transcribed. Non-sense mutations almost always lead to a non-functioning protein and are therefore the most serious mutations.

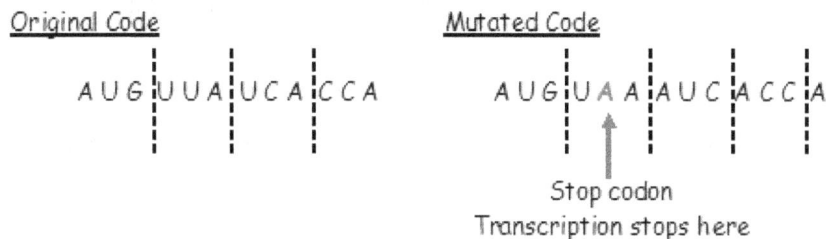

Original Code   Mutated Code

A U G ¦U U A ¦U C A ¦C C A   A U G ¦U A A ¦A U C ¦A C C ¦A

Stop codon
Transcription stops here

Fig.19.4

Cystic Fibrosis

Cystic fibrosis results from mutations in the CFTR gene found on chromosome 7. It is 250,000 base pairs long, and creates a protein that is 1,480 amino acids long.

- The most common mutation, is a **deletion** of three nucleotides that results in a loss of the amino acid phenylalanine at the 508th (508) position on the protein. This mutation accounts for two-thirds of CF cases worldwide, however, there are over 300 other mutations that can produce CF.
- One mutation creates a protein that does not fold normally and is degraded by the cell.
- Several mutations result in proteins that are too short because production is ended prematurely by a stop codon.
- Less common mutations produce proteins that do not use energy normally, do not allow chloride to cross the membrane appropriately, or are degraded at a faster rate than normal.
- Mutations may also lead to fewer copies of the CFTR protein being produced.

Consequenses of large number of alleles

Each mutation results in a slightly different form of the CFTR gene. These forms of the genes are called as alleles. Each allele will express itself in different ways. So, the symptoms of the disorder will be highly variable. It also means that genetic testing for the disorder will be more difficult as all the alleles will have to be identified.

CHAPTER TWENTY
MONOHYBRID INHERITANCE

> Learning outcomes: by the end of this chapter you should be able to
> **Edexcel Syllabus Spec 16:** *Explain the terms gene, allele, genotype, phenotype, recessive, dominant, homozygote and heterozygote, and explain monohybrid inheritance, including the interpretation of genetic pedigree diagrams, in the context of traits such as cystic fibrosis, albinism, thalassaemia, garden pea height and seed morphology.*

Gene
A gene is a sequence of bases on DNA that codes for a single polypeptide chain. The gene locus is the location of a gene on the chromosome.

Allele
A gene can exist in two or more different forms. These **alternate forms of a gene found at a particular gene locus are referred to as alleles**. The alleles differ from each other in the sequence of bases.

It is conventional to represent the dominant allele by a capital letter and the corresponding recessive allele by the simple form of the same letter. For e.g.: The gene for **height in pea plants** can exist in two forms – **T** (allele for tallness) or **t** (allele for dwarfism) thus the alleles of the gene for height are **T** and **t**.

Genotype
The combination of alleles found at the same gene locus of homologous chromosomes. For example the genotypes TT and tt.

Homozygote
A genotype consisting of similar alleles. For example the genotypes TT and tt are homozygous.

Heterozygote
A genotype consisting of two different alleles. For example the genotypes Tt is heterozygous.

Dominant allele
The allele which expresses itself in the phenotype, both, in homozygous and in heterozygous genotypes. For example the allele T expresses itself when found as TT to give rise to tall pea plants and also when found as Tt it produces tall pea plants.

Recessive allele
The allele which expresses itself in the phenotype, only in homozygous conditions, but, not in heterozygous conditions. For example the allele t produces dwarf pea plants on in tt condition. In heterozygous condition (Tt), the recessive allele(t) is present, but is not expressed in the phenotype.

Phenotype
The observable features of an organism that are produced by the interaction of the genotype and the environment. For example Tallness or Dwarfism are the observable features of pea plants.

Monohybrid cross
The monohybrid cross involves the inheritance of a single pair of genes. The gene has two possible alleles. One allele is dominant and the other is recessive.

Consider an example of a monohybrid cross involving the inheritance of shape of pea seeds, illustrated in fig.20.1. This gene has two alleles - Round (**R**) and Wrinkled (**r**). It is conventional to represent the dominant allele by a capital letter and the corresponding recessive allele by the simple form of the same letter.

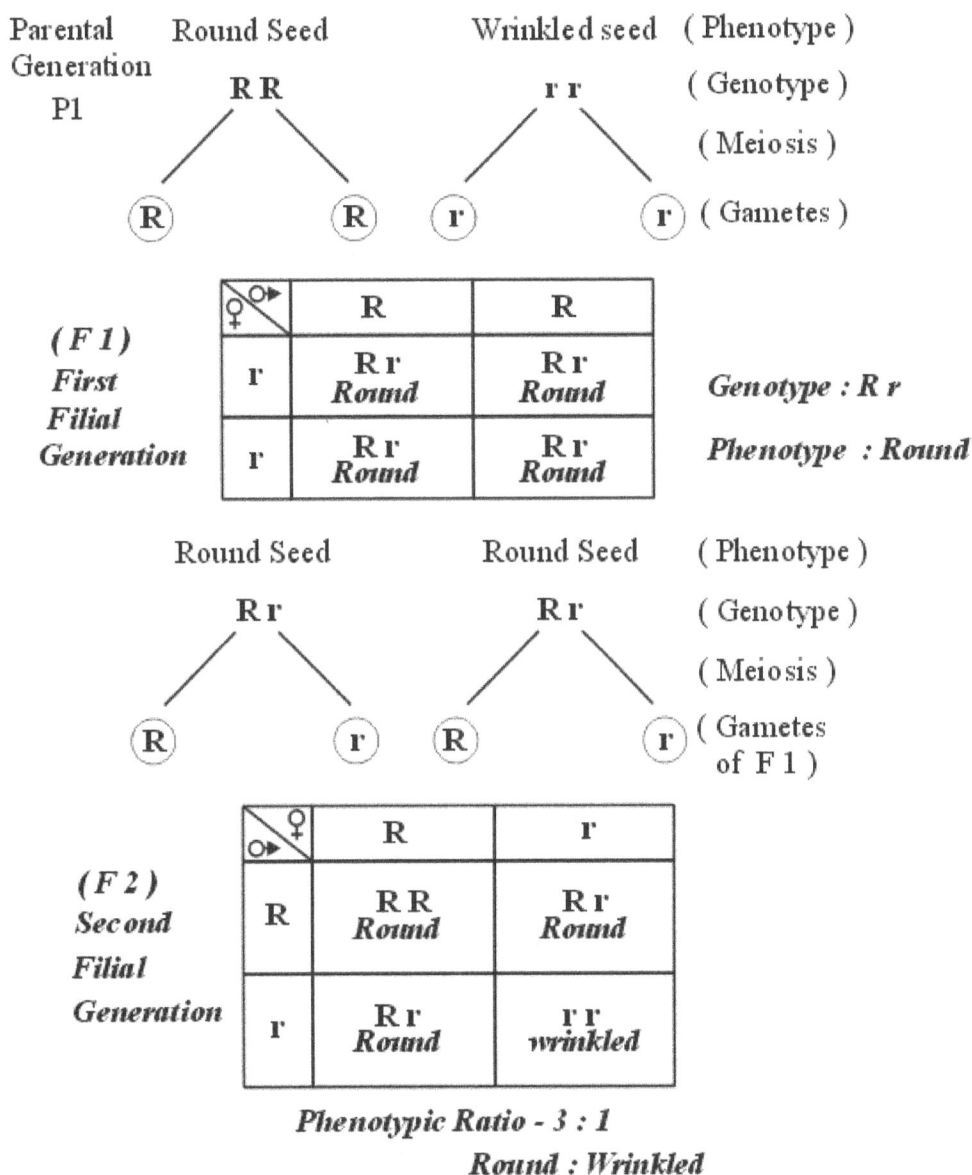

Parental Generation P1 — Round Seed **RR** — Wrinkled seed **rr** (Phenotype) (Genotype) (Meiosis)

Gametes: R, R, r, r

(F 1) First Filial Generation

♀\♂	R	R
r	R r *Round*	R r *Round*
r	R r *Round*	R r *Round*

Genotype : *R r*

Phenotype : *Round*

Round Seed **Rr** — Round Seed **Rr** (Phenotype) (Genotype) (Meiosis)

Gametes of F 1: R, r, R, r

(F 2) Second Filial Generation

♂\♀	R	r
R	R R *Round*	R r *Round*
r	R r *Round*	r r *wrinkled*

Phenotypic Ratio - 3 : 1

Round : Wrinkled

Fig.20.1

The Genetics of autosomal recessive disorders

Humans have 23 pairs of chromosomes – 22 pairs of **autosomes** (which do not play a role in sex determination) and 1 pair of **sex chromosomes** (which play a role in sex determination). **Autosomal disorders** are caused by genes on the autosomes and can affect both males and females indiscriminately (equally). Disorders caused by gene located on the sex chromosomes are called as sex – linked disorders. **Recessive** means that it is caused by a recessive allele and can cause the disorder **only if found in homozygous condition. Cystic fibrosis, Thalassaemia** and **albinism** are all autosomal recessive disorders.

All people have the same genes, which code for the proteins that make up our body. Because we have 2 copies of each chromosome we have 2 copies of each gene. However, it is possible to have different versions of a gene. The versions are called **alleles**. Because alleles are different versions of the **same gene** they are found in the **same position** on each of the pairs of chromosomes. The position that a gene occupies on a chromosome is called a **Locus** (latin = place).

Alleles at a particular locus can be either **recessive** or **dominant**
Dominant: Always affects the phenotype

Recessive: Does not affect the phenotype if a dominant allele is present

Different alleles arise through **mutation**. Mutations that cause non-functional alleles to develop can lead to **genetic diseases, as shown in fig.20.2.**

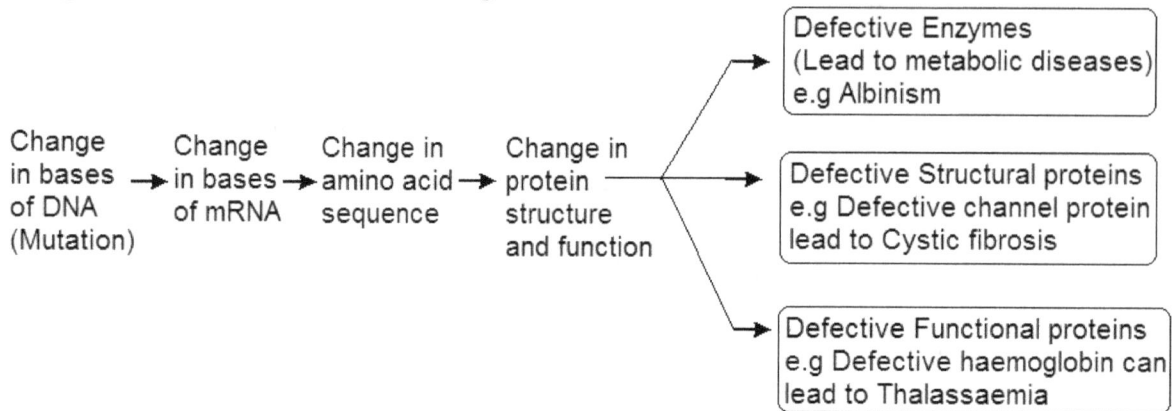

Change in bases of DNA (Mutation) → Change in bases of mRNA → Change in amino acid sequence → Change in protein structure and function

→ Defective Enzymes (Lead to metabolic diseases) e.g Albinism

→ Defective Structural proteins e.g Defective channel protein lead to Cystic fibrosis

→ Defective Functional proteins e.g Defective haemoglobin can lead to Thalassaemia

Fig.20.2

Usually a genetic disorder is recessive, this means you **have to inherit two copies of the mutant allele** to cause the genetic disease. The genotype and corresponding phenotypes for recessive disorders is shown below:

AA: Homozygous dominant. Phenotype is normal. Both genes produce normal proteins.

Aa: Heterozygous (Genetic carrier). Phenotype is normal. One gene produces the defective protein and the other gene produces the normal protein. However, there is usually enough of the normal protein to carry out the normal function. The phenotype is normal but the defective gene can be passed on to offspring..

aa: Homozygous recessive. Both genes produce the defective protein. So, the individual will be suffering from the genetic disorder.

Disease	Heritability	Effect
Cystic fibrosis	Recessive	Defective CFTR protein in epithelial cell membrane. Mucus thick and sticky due to reabsorption of water into epithelial cells. Sweat is also very salty. Respiratory system, digestive system and reproductive system are affected.
Thalassaemia	Recessive	Reduced rate of synthesis of one of the globin chains that make up haemoglobin or formation of defective haemoglobin. Reduced synthesis or defective globin chains can cause the formation of abnormal hemoglobin molecules, and this in turn causes the anaemia which is the characteristic symptom of the thalassaemics.
Albinism	Recessive	Albinism is caused by a mutant allele, which prevents the synthesis of the skin pigment melanin. It results from a genetic defect in an enzyme called tyrosinase. This enzyme helps the body to change the amino acid tyrosine into a pigment called melanin. Inactive tyrosinase leads to white hair and very light skin.

The diagram below shows the inheritance of single gene traits involving both autosomal recessive and autosomal dominant disorders in pedigree charts.

symbols used in family trees and genetic diagrams

☐ = male   ○ = female

☐ ○ unaffected

■ ● affected

◧ ◑ heterozygous carrier (autosomal recessive)

☐═○ indicates consanguineous couple (parents of offspring closely related)

A = dominant allele
a = recessive allele

autosomal recessive disorders

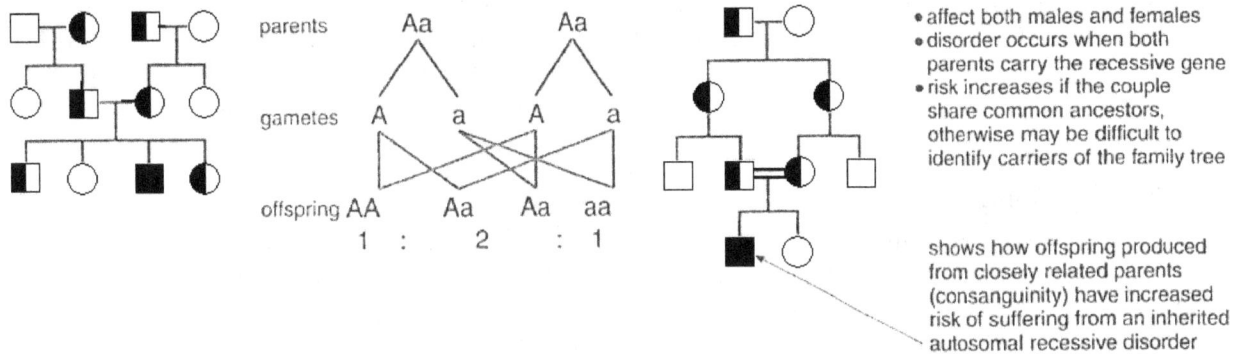

parents   Aa   Aa

gametes   A   a   A   a

offspring   AA   Aa   Aa   aa
             1   :   2   :   1

• affect both males and females
• disorder occurs when both parents carry the recessive gene
• risk increases if the couple share common ancestors, otherwise may be difficult to identify carriers of the family tree

shows how offspring produced from closely related parents (consanguinity) have increased risk of suffering from an inherited autosomal recessive disorder

Fig.20.3

Some rules for pedigree charts

- Individuals in the same generation are placed in the same row. For example, Vazna, Amir, Vulau and Shifa all belong to the same generation. Elaf, Sophia, Fizal and Faraha belong to the next generation.
- The older offspring is placed to the right of the younger offspring in each family. For example, Vazna is younger than Amir and Vulau is younger than Shifa.

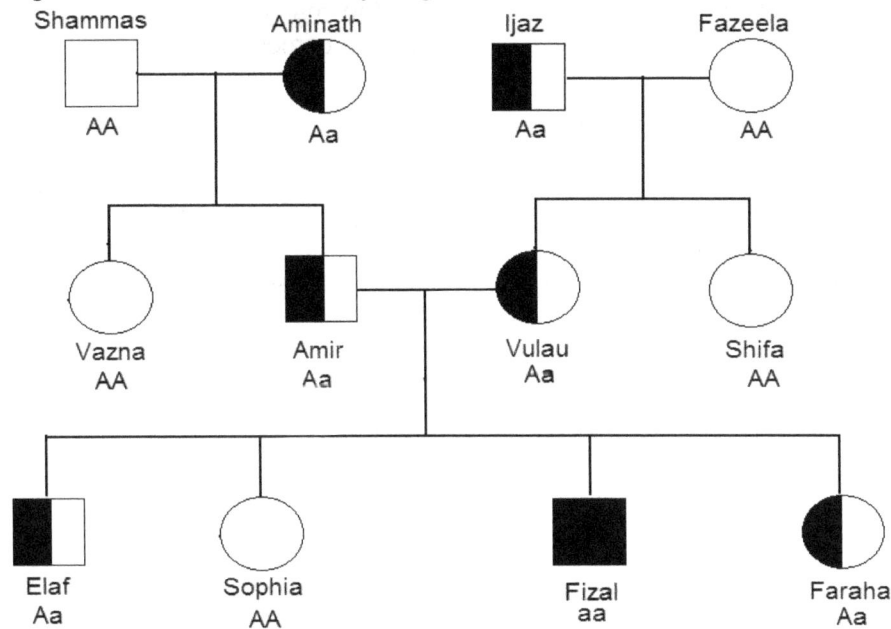

Shammas — Aminath   Ijaz — Fazeela
AA   Aa   Aa   AA

Vazna   Amir   Vulau   Shifa
AA   Aa   Aa   AA

Elaf   Sophia   Fizal   Faraha
Aa   AA   aa   Aa

Fig.20.4

Examples of inheritance:

Vazna (**AA**) has inherited one **A** allele from each parent. So, Vazna will be normal.

Amir (**Aa**) has inherited one dominant (**A**) allele from Shammas and one recessive (**a**) allele from Aminath. So, Amir is a genetic carrier and can pass on the recessive (disease causing allele) allele to his offspring (Fizal and Elaf). But, Amir is phenotypically normal.

Fizal (**aa**) has inherited one recessive (**a**) allele from each parent (Amir and Vulau). So, Fizal suffers from a genetic disorder caused by recessive alleles.

CHAPTER TWENTY ONE
CYSTIC FIBROSIS

Learning outcomes: by the end of this chapter you should be able to
Edexcel Syllabus Spec 17: *Explain how the expression of a gene mutation in people with cystic fibrosis impairs the functioning of the gaseous exchange, digestive and reproductive systems.*

Cystic fibrosis (CF) is caused by a defect in a gene called the cystic fibrosis transmembrane conductance regulator (CFTR) gene. This gene makes a protein that controls the movement of salt and water in and out of the cells in your body. In people with Cystic Fibrosis, the CFTR protein does not work effectively. This causes the thick, sticky mucus and very salty sweat, which are the main features of CF.

The functioning of the CFTR protein is illustrated in the fig.21.1, 21.2 and 21.3

Normal Situation **Excess water in mucus**	**Normal situation** **Dehydrated mucus - less water**
1. Na^+ ions are actively pumped across the basal membrane, into tissue fluid. 2. Na^+ diffuses from mucus into the cells, through sodium channels in the apical membrane. 3. Cl^- diffuses down electrical gradient. 4. Water is drawn out of cells at the basal end, by osmosis, due to the high salt concentration in the tissue fluid. 5. Water is drawn out of the mucus by osmosis. So, mucus becomes more viscous.	1. Cl^- ions are pumped into the cell across the basal membrane. 2, The CFTR channel opens and Cl^- ions diffuse out of the cell, into the mucus (secretion of Cl^- ions). 3. The CFTR channel closes the Na^+ channels, but Na^+ ions diffuse down the electrical gradient into the mucus. 4. The increased concentration of Na^+ and Cl^- ions in the mucus results in water entering the mucus, by osmosis. So mucus becomes more runny (less viscous).

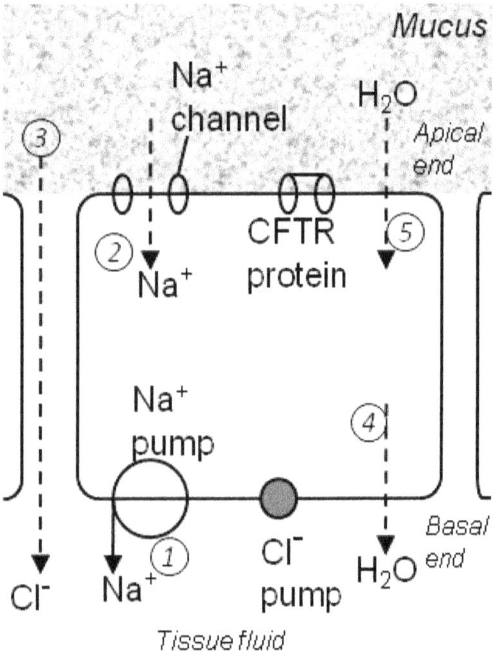

Fig.21.1

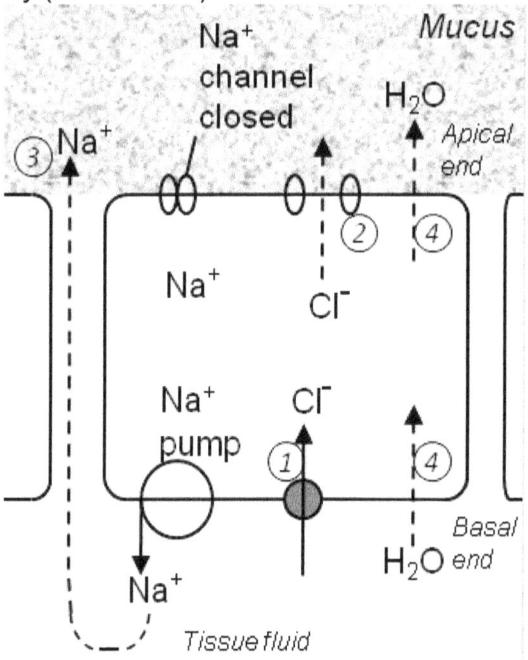

Fig.21.2

Cystic fibrosis situation (CFTR – Mutated)	
1. CFTR channel is absent or dysfunctional and is not able to regulate the sodium channel. 2. **Na+ channel is permanently open** and Na+ ions continue to enter the cell at the apical end, through the open Na+ channels. 3. No Cl⁻ secretion takes place. 4. Water is continually removed from mucus by osmosis. **Mucus is always thick and sticky.**	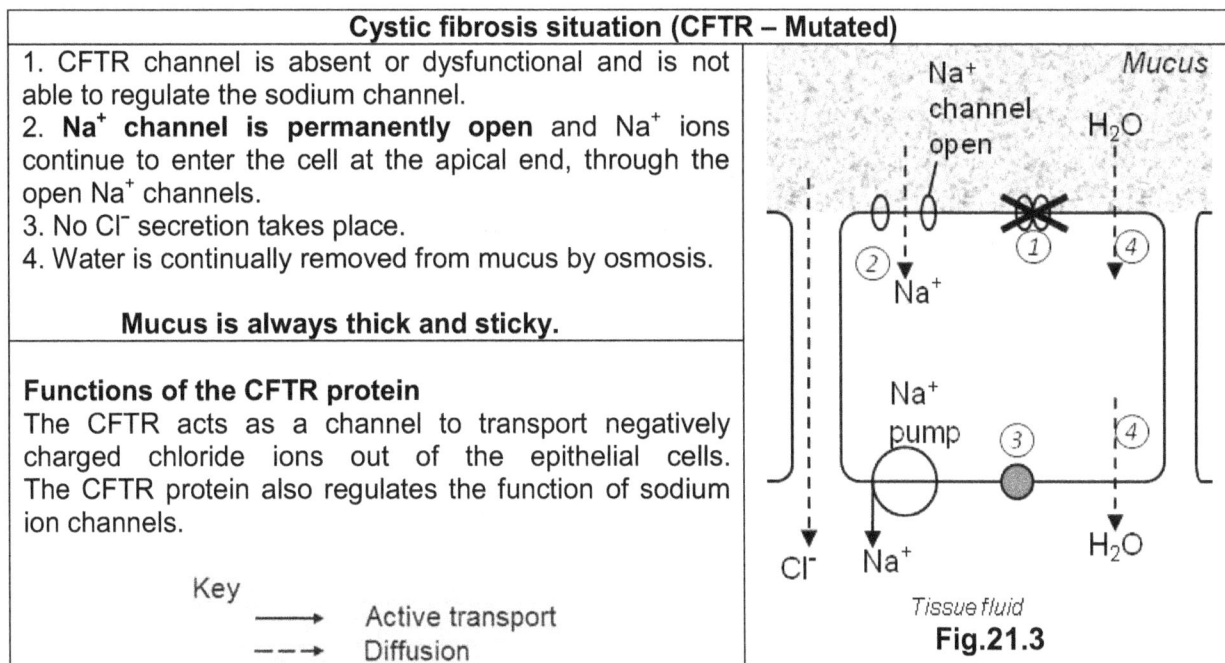

Functions of the CFTR protein

The CFTR acts as a channel to transport negatively charged chloride ions out of the epithelial cells. The CFTR protein also regulates the function of sodium ion channels.

Key

⟶   Active transport

--→   Diffusion

Fig.21.3

The effects of cystic fibrosis on different systems are described below.

Digestive system

The thick, sticky mucus can also block tubes, or ducts, in the pancreas. As a result, digestive enzymes that are produced by the pancreas cannot reach the small intestine. These enzymes help break down the food. Without them, the intestines cannot absorb fats and proteins fully.
As a result:
- Nutrients leave the body unused, and the individual becomes malnourished.
- Stools become bulky. Some newborns require surgery to remove the bulky stools.
- There may be intestinal gas, a swollen belly, and pain or discomfort.
- Pancreatitis. Pancreatitis is inflammation in the pancreas that causes pain.
- Rectal prolapse. Frequent coughing or problems passing stools may cause rectal tissue from inside to move out of the rectum.
- Liver disease due to inflammation or blocked bile ducts.

Sweat glands

The abnormal gene also causes the sweat to become extremely salty. As a result, the body loses large amounts of salt. This can upset the balance of minerals in the blood.

Lungs

The lungs are normal at birth, but breathing problems can develop at any time afterward. Thick secretions eventually block the small airways, which leads to **inflammation and thickening** of their walls. As larger airways fill with secretions, areas of the lung collapse and contract and the lymph nodes enlarge. All these changes make breathing increasingly difficult and reduce the lungs' ability to transfer oxygen to the blood. Respiratory tract infections occur because of bacterial growth in the bronchial secretions and walls of the airways.

Bronchiectasis is a lung disease in which the bronchial tubes, or large airways in your lungs, become stretched out and flabby over time and form pockets where mucus collects. The mucus provides a breeding ground for bacteria. This leads to repeated lung infections. If not treated, bronchiectasis can lead to serious illness, including respiratory failure. Recent research indicates that the extracellular chloride also contributes to infection by disabling a natural antibiotic made by some body cells. When immune cells come to the rescue, their remains add to the mucus, creating a vicious cycle.

Reproductive system

People with cystic fibrosis often have impaired reproductive function. Almost all men have a low sperm count (which makes them sterile) because one of the ducts of the testis (the vas deferens) has developed abnormally and blocks the passage of sperm.

In women, cervical secretions are too thick, causing decreased fertility. Otherwise, sexual function is not affected. Women with cystic fibrosis have a higher likelihood of complications during pregnancy (such as developing a lung infection or diabetes), but many women with cystic fibrosis have given birth.

The effects of cystic fibrosis are summarised in fig.21.4.

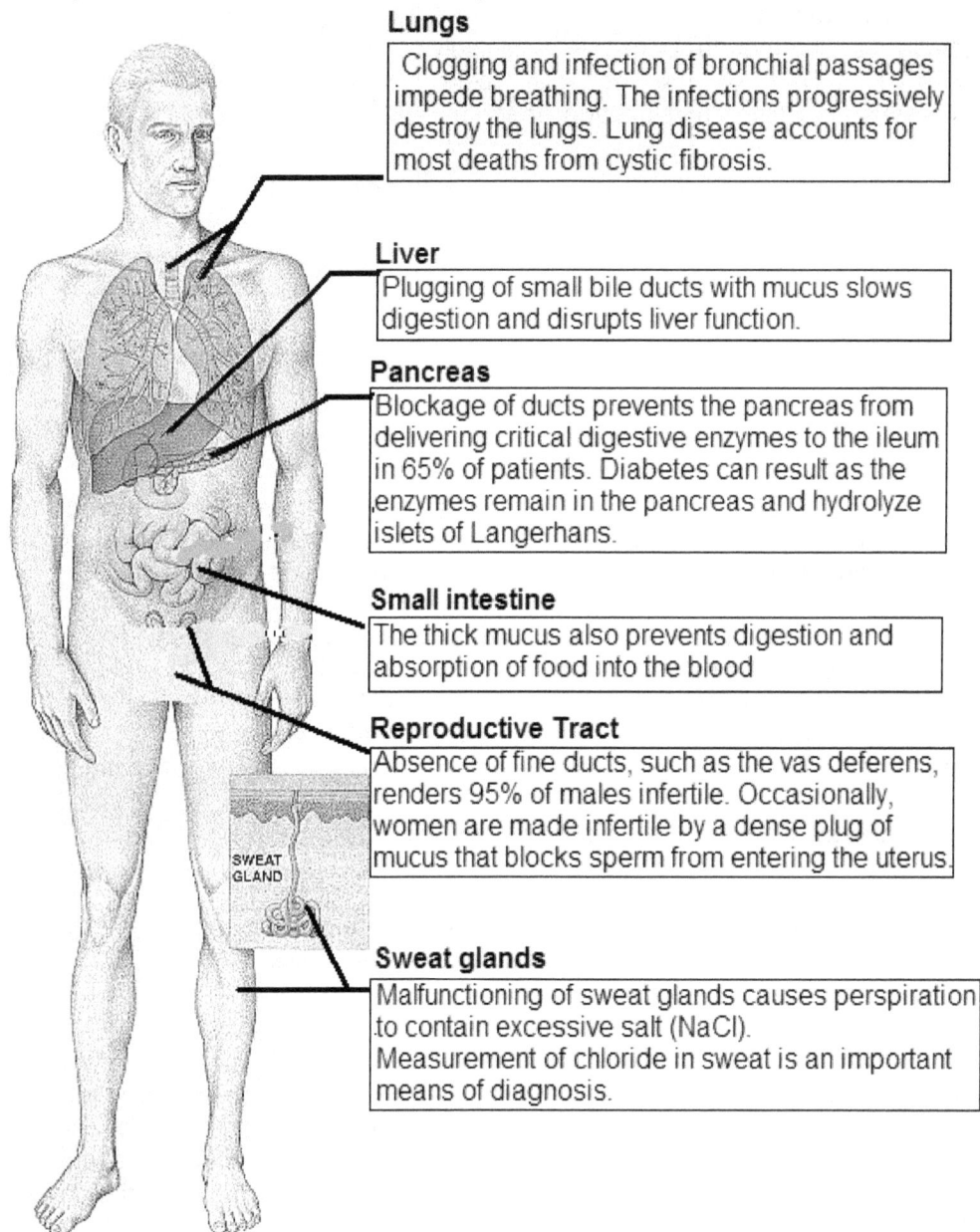

Lungs

Clogging and infection of bronchial passages impede breathing. The infections progressively destroy the lungs. Lung disease accounts for most deaths from cystic fibrosis.

Liver

Plugging of small bile ducts with mucus slows digestion and disrupts liver function.

Pancreas

Blockage of ducts prevents the pancreas from delivering critical digestive enzymes to the ileum in 65% of patients. Diabetes can result as the enzymes remain in the pancreas and hydrolyze islets of Langerhans.

Small intestine

The thick mucus also prevents digestion and absorption of food into the blood

Reproductive Tract

Absence of fine ducts, such as the vas deferens, renders 95% of males infertile. Occasionally, women are made infertile by a dense plug of mucus that blocks sperm from entering the uterus.

Sweat glands

Malfunctioning of sweat glands causes perspiration to contain excessive salt (NaCl).
Measurement of chloride in sweat is an important means of diagnosis.

SWEAT GLAND

Fig.21.4

CHAPTER TWENTY TWO
GENE THERAPY

Learning outcomes: by the end of this chapter you should be able to
Edexcel Syllabus Spec 18: *Describe the principles of gene therapy and distinguish between somatic and germ line therapy.*
Edexcel Syllabus Spec 19: *Explain the uses of genetic screening: identification of carriers, preimplantation genetic diagnosis and prenatal testing (amniocentesis and chorionic villus sampling) and discuss the implications of prenatal genetic screening.*
Edexcel Syllabus Spec 20: *Identify and discuss the social and ethical issues related to genetic screening from a range of ethical viewpoints.*

The idea of gene therapy is to **genetically alter human cells** in order to **treat a genetic disorder**. Gene therapy may involve altering the **genotype of a tissue (Somatic gene therapy)** or even a **whole human (Germ line therapy).**

Somatic cell gene therapy
Somatic cell gene therapy means genetically altering **specific body (or somatic) cells**, such as bone marrow cells, pancreas cells, or whatever, in order to treat the disease. This therapy may treat or cure the disease, but any genetic **changes will not be passed on their offspring.**

Germ-line therapy
Germ-line therapy means genetically altering those cells (**sperm cells, sperm precursor cell, ova, ova precursor cells, zygotes or early embryos**) that will pass their genes down the "germ-line" to future generations. Alterations to any of these cells will affect every cell in the resulting human, and in all his or her descendants.

Germ-line therapy would be highly effective, but is also potentially **dangerous** (since the long-term effects of genetic alterations are not known), **unethical** (since it could easily lead to eugenics – alteration of human genome for improving the qualities of the human species or human population) and **immoral** (since it could involve altering and destroying human embryos). Germ line gene therapy is currently illegal in the UK and most other countries, and current research is focusing on somatic cell therapy only.

The flowchart below outlines the principles of gene therapy

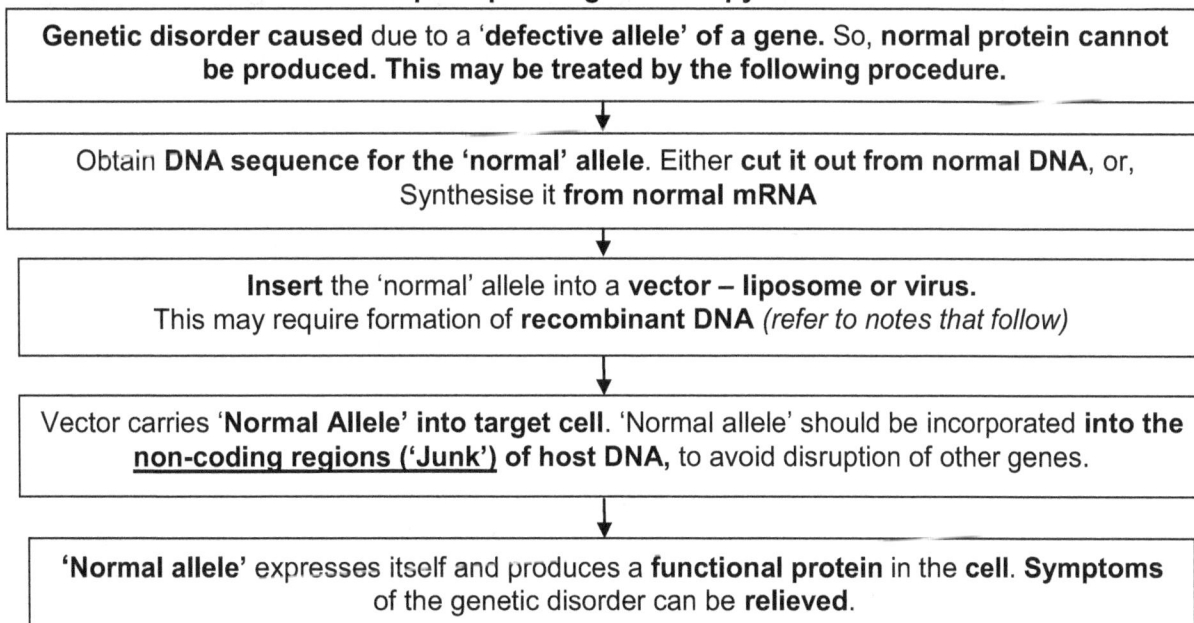

Genetic disorder caused due to a **'defective allele'** of a gene. So, **normal protein cannot be produced. This may be treated by the following procedure.**

↓

Obtain **DNA sequence for the 'normal' allele**. Either **cut it out from normal DNA**, or, Synthesise it **from normal mRNA**

↓

Insert the 'normal' allele into a **vector – liposome or virus.** This may require formation of **recombinant DNA** *(refer to notes that follow)*

↓

Vector carries '**Normal Allele**' into target cell. 'Normal allele' should be incorporated **into the non-coding regions ('Junk') of host DNA,** to avoid disruption of other genes.

↓

'Normal allele' expresses itself and produces a **functional protein** in the **cell. Symptoms** of the genetic disorder can be **relieved.**

A few **tools** and **techniques** used in gene therapy are illustrated below.

1. Obtaining the normal allele or gene
a. Synthesis of the normal allele by using normal mRNA

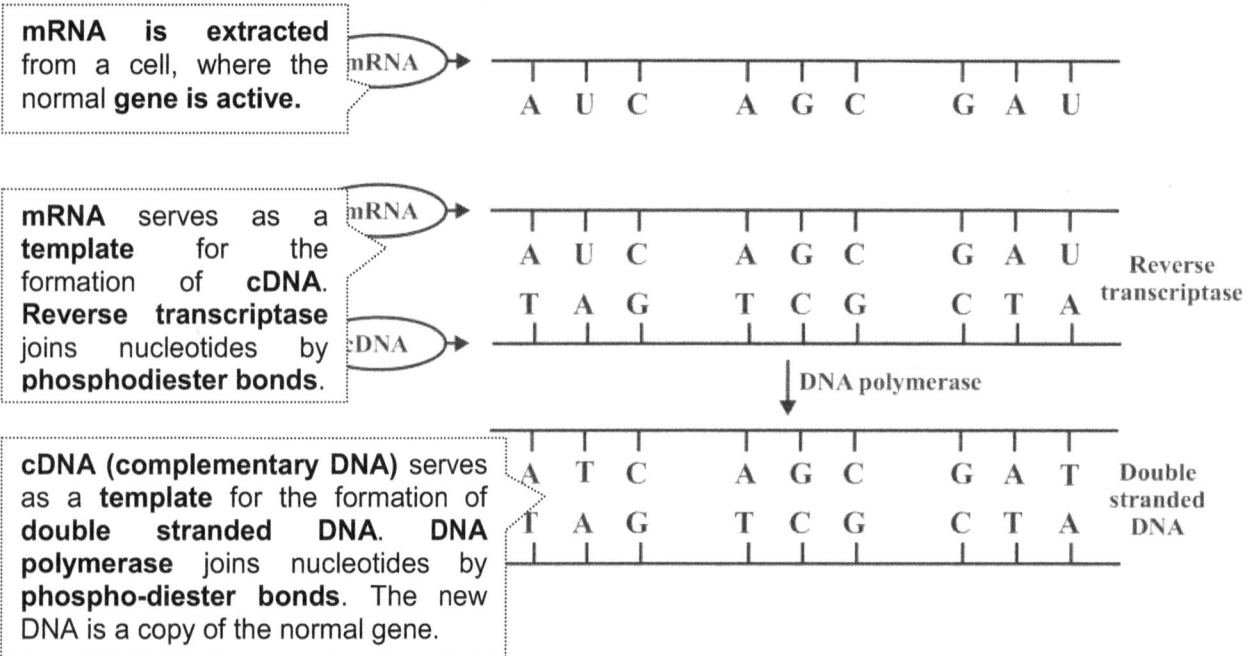

mRNA is extracted from a cell, where the normal **gene is active.**	mRNA →

A U C   A G C   G A U

mRNA serves as a **template** for the formation of **cDNA**. **Reverse transcriptase** joins nucleotides by **phosphodiester bonds.**	mRNA → / cDNA →

A U C   A G C   G A U   *Reverse*
T A G   T C G   C T A   *transcriptase*

↓ DNA polymerase

cDNA (complementary DNA) serves as a **template** for the formation of **double stranded DNA. DNA polymerase** joins nucleotides by **phospho-diester bonds**. The new DNA is a copy of the normal gene.	

A T C   A G C   G A T   *Double*
T A G   T C G   C T A   *stranded DNA*

Fig.22.1

b. Isolation of the 'normal allele' from a DNA molecule by using restriction endonucleases

<u>Restriction Endonucleases</u> are enzymes that cut DNA into small fragments. This allows individual genes to be isolated by cutting specific regions of DNA on either side of the desired gene.

The restriction endonuclease makes a staggered cut in the DNA molecule between specific base sequences. There are hundreds of different restriction endonucleases, each cuts DNA at a specific base sequence. Example of the restriction endonucleases are ECORI and HIND III will always cut DNA between the bases as shown.

The restriction endonuclease ECORI will always cut DNA between the bases as shown below:

G | A A T C   G | A A T C
C T T A A | G   C T T A A | G

The restriction endonuclease **Hind III** will cut DNA between the base sequences as shown below

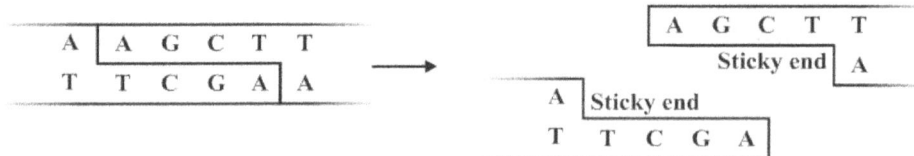

A | A G C T T   A G C T T
T T C G A | A   Sticky end | A

                       A | Sticky end
                       T T C G A

Fig.22.2

The specific nucleotide sequence between which the endonuclease cuts the DNA is called a **recognition site or restriction site**. The cut ends are complementary to each other and will have a natural affinity for each other. These are referred to as 'Sticky ends'. Some restriction endonucleases also cut DNA to form **blunt ends.**

2. Pasting DNA fragments together (Formation of recombinant DNA)

Once the donor DNA and recipient DNA have been cut by the **same restriction endonuclease**, the sticky ends are joined up using the enzyme **DNA ligase**. This helps to form **Recombinant DNA.**

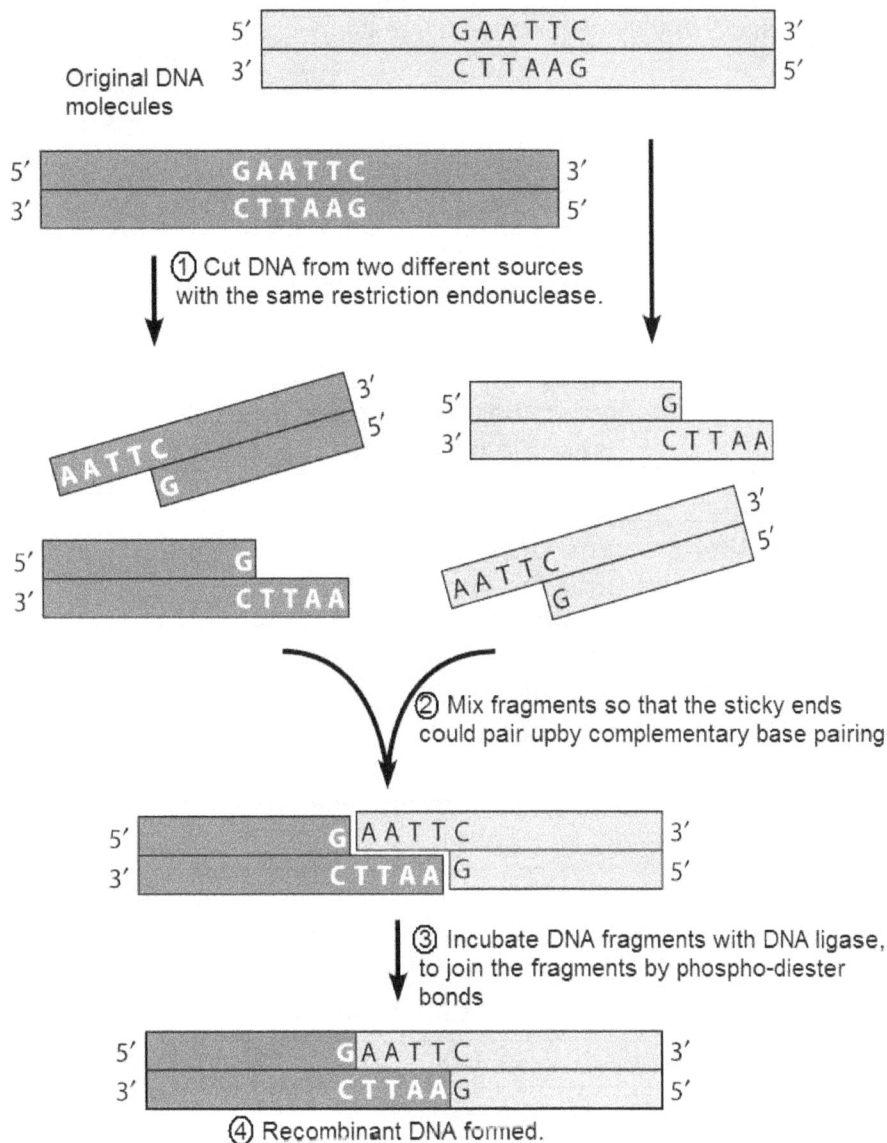

Fig.22.3

Note: Plasmids are circular loops of extra-chromosomal DNA in bacterial cells. The small size of plasmids and their ability to replicate within cells, makes them good genetic tools to transfer genes into cells. Recombinant DNA may involve the use of plasmids.

3. Inserting Recombinant DNA into host cells.

Host cells that take up the recombinant DNA are called **transformed cells or transgenic cells.** Vectors containing normal allele must be incorporated into living cells so that they can be replicated or expressed. The cells receiving the normal allele are called <u>host cells</u>, and once they have successfully incorporated the recombinant DNA they are said to be <u>transformed</u>.

DNA and genes are large molecules which do not readily cross cell membranes, so the membranes must be made permeable in some way. There are different ways of doing this depending on the type of host cell.

This can be achieved by one of the methods shown in fig.22.4 or 22.5.

a. Inserting genes into cells by using viruses

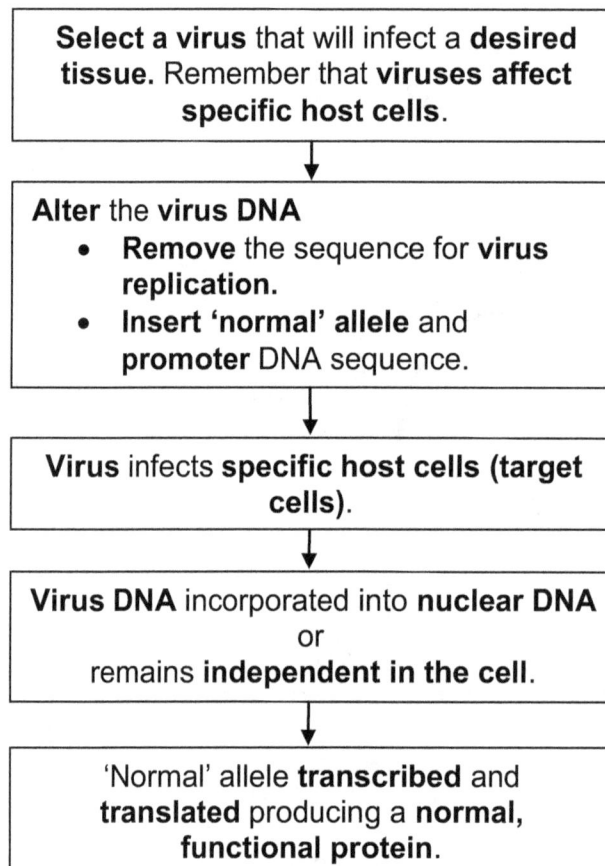

Select a virus that will infect a **desired tissue.** Remember that **viruses affect specific host cells**.

↓

Alter the **virus DNA**
- **Remove** the sequence for **virus replication.**
- **Insert 'normal' allele** and **promoter** DNA sequence.

↓

Virus infects **specific host cells (target cells).**

↓

Virus DNA incorporated into **nuclear DNA**
or
remains **independent in the cell**.

↓

'Normal' allele **transcribed** and **translated** producing a **normal, functional protein.**

Fig.22.4

b. Inserting genes into cells by using liposomes

Liposomes are small hollow spheres made up of a phospholipid bilayer. 'Normal alleles' can be encased in liposomes, which are small membrane vesicles. The liposomes fuse with the cell membrane (and sometimes the nuclear membrane too), delivering the DNA into the cell. This works for many types of cells, but is particularly useful for delivering genes to cell *in vivo* (such as in gene therapy).

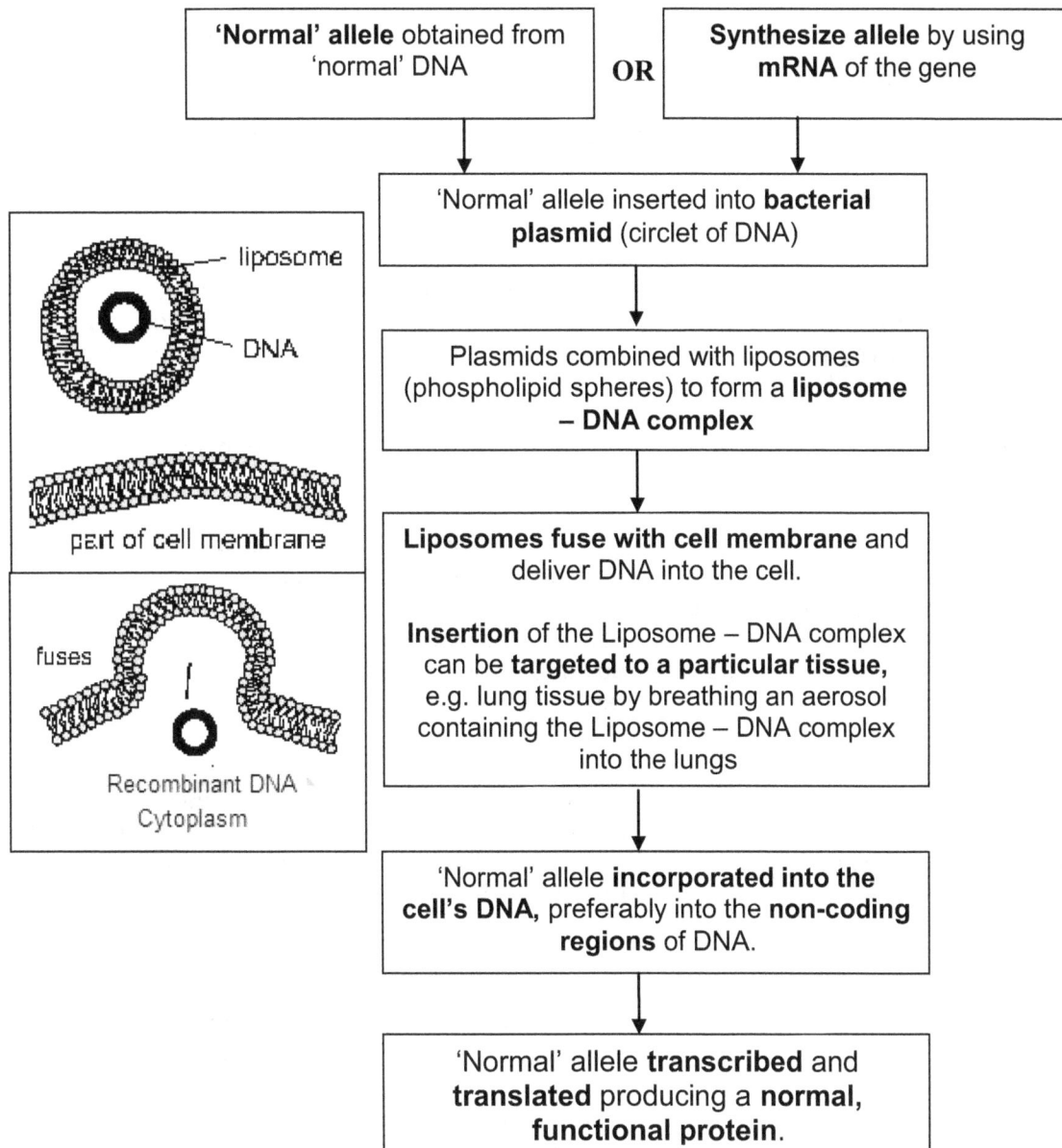

'Normal' allele obtained from 'normal' DNA	OR	**Synthesize allele** by using **mRNA** of the gene

'Normal' allele inserted into **bacterial plasmid** (circlet of DNA)

Plasmids combined with liposomes (phospholipid spheres) to form a **liposome – DNA complex**

Liposomes fuse with cell membrane and deliver DNA into the cell.

Insertion of the Liposome – DNA complex can be **targeted to a particular tissue**, e.g. lung tissue by breathing an aerosol containing the Liposome – DNA complex into the lungs

'Normal' allele **incorporated into the cell's DNA**, preferably into the **non-coding regions** of DNA.

'Normal' allele **transcribed** and **translated** producing a **normal, functional protein**.

Fig.22.5

The 'normal' gene must be incorporated **into the non-coding region (junk)** of the DNA. If the new section of DNA is inserted **into a sequence of DNA that is already coding for a protein** it will **disrupt the original coding sequence** and will most likely produce a **non functional protein**.

Gene therapy and cystic fibrosis

Cystic fibrosis (CF) is the most common genetic disease in the UK, affecting about 1 in 2500. CF is always fatal, though life expectancy has increased from 1 year to about 20 years due to modern treatments.

These treatments include physiotherapy many times each day to dislodge mucus from the lungs; antibiotics to fight infections; DNAse drugs to loosen the mucus; enzymes to help food digestion and even a heart-lung transplant.

The complicated and ultimately unsuccessful treatment, make CF as **a good candidate for gene therapy** and is one of the first disease to be treated in this way.

The gene for CFTR was identified in 1989 and a cDNA clone was made soon after. The idea is to deliver copies of this good gene to the epithelial cells of the lung, where they can be incorporated into the nuclear DNA and make functional CFTR chloride channels. If about 10% of the cells could be corrected, this would cure the disease.

Recent research has shown that approximately 20% of the epithelial cells of cystic fibrosis patients accepted the CFTR gene. However the effects are **short-lived** as the epithelial cells are constantly replaced at a steady rate. So, **regular treatment** with liposome-DNA complex aerosols is needed to provide long term improvement of the symptoms.

Germ line genetic engineering is still in its infancy. Already, however, it has been important for producing a variety of types of specially altered animals.

Benefits of Gene Therapy	Risks of Gene Therapy
Gene therapy could be used to prevent or **cure genetic disorders**, but this is only likely **if germ line therapy is allowed**. This is currently not permitted due to ethical concerns over possible effects on future generations.	Viruses are commonly used as vectors. These viruses might mutate to become harmful and cause other diseases. Some viruses may cause adverse and sometimes fatal immune responses.
Genetic disorders can be **treated** by **replacing faulty genes with functioning genes** in **somatic cells**.	The gene should be **incorporated into the non-coding regions**, called 'Junk'. If the gene gets inserted in between another gene, by splitting the base sequences, it could cause harm. This has been shown to happen in some patients, resulting in leukaemia.
Gene therapy has the potential to treat a variety of disorders including thalassaemia and cystic fibrosis	Gene therapy is a fairly new technique and long term effects are not yet known. Effects of somatic cell therapy are likely to be temporary as cells have a limited life span. Currently only somatic gene therapy is allowed, so it is difficult to target large areas of the body.

Genetic screening is the **detection of mutant genes** by analysis of DNA, using radioactive probes.

Uses of genetic screening
1. Identification of carriers
Since most children with **recessive disorders** are born to parents with **normal phenotypes,** genetic screening is used to determine whether or not a **prospective parent** is a carrier. This is useful for parents to make informed decisions about having children. In some countries, premarital testing is encouraged to reduce the incidence of heterozygous couples.

2. Detection of defective embryos
a. Pre-implantation Genetic diagnosis
Two carriers of a genetic disorder donate sperm and ova. *In* vitro fertilization is carried out and the embryos are allowed to grow to the eight cell stage in a Petri dish. A single cell is extracted from each embryo and genetic screening is carried out. Healthy embryos are identified and implanted into the uterus of the mother. This technique allows two carriers to have a healthy normal baby.

b. Post-implantation Genetic diagnosis or prenatal testing
Prenatal testing is used to determine whether or not an embryo has a disease. It helps the parents to make informed decisions about termination of the pregnancy by abortion, or gives them enough time to make preparations to deal with the arrival of a defective child. It also allows physicians to start treatment at the earliest.

i. Amniocentesis	ii. Chorionic Villus Sampling
A sample of amniotic fluid containing fetal cells is withdrawn through the abdominal wall at about 15 to 16 weeks' gestation. The amniotic cells can be cultured for further tests. Analysis of DNA is carried out to identify specific gene defects. The results from cultured cells are available in about 3 to 4 weeks. The test is reliable and safe with low risk of miscarriage	Cells are taken from the chorionic villi, through the cervix or through abdominal wall, under ultrasonic guidance. Biochemical and chromosomal tests are carried out on fetal cells. The culturing of cells is not needed, so the results are obtained more quickly than amniocentesis. It is performed at between 9 and 12 weeks gestation. The risk of miscarriage is slightly higher than normal at this time

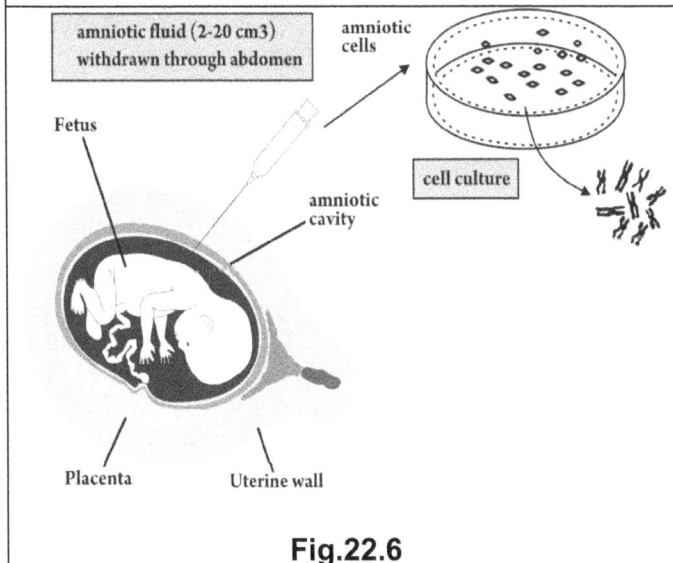

amniotic fluid (2-20 cm3) withdrawn through abdomen

amniotic cells

Fetus

cell culture

amniotic cavity

Placenta   Uterine wall

Fig.22.6

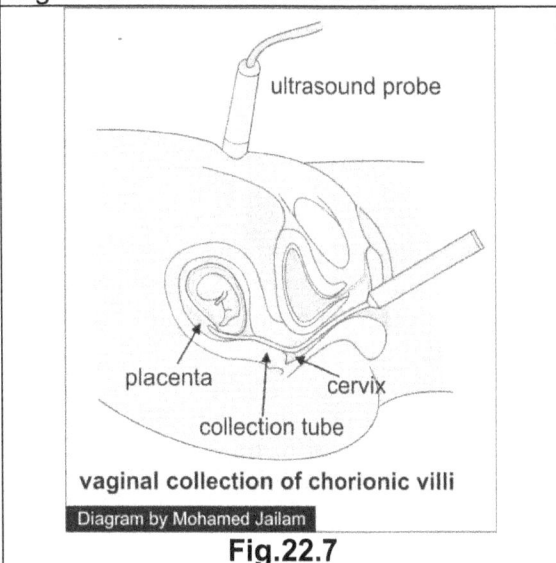

ultrasound probe

placenta   cervix

collection tube

vaginal collection of chorionic villi

Diagram by Mohamed Jailam

Fig.22.7

Social and ethical views **on prenatal screening**

Against genetic screening	In **favour** of genetic screening
Abortion and miscarriage There are ethical and social issues associated with genetic screening of unborn fetuses because this procedure allows parents to **abort** fetuses with a chromosome or gene abnormality. There is also evidence that, in some parts of the world, **abortion** on the basis of **gender** is carried out. This is a social issue.	**Economical** A single treatment by germ line gene therapy is preferable to continuous, expensive medical care. For example, cystic fibrosis could cost a patient up to an average of $46,500 per year and $146, 500 over a lifetime. However, germ line gene therapy is still not permissible.
Decision making In Cyprus, the screening programme for thalassemia lead to social pressures on couples to abort fetuses that they did not personally want to abort. A mother has the right to continue or not continue with a pregnancy, particularly if the pregnancy presents risks to the mother's physical or mental health. Parents have the right to make a free and properly informed choice about whether or not to abort a fetus found to have a significant disorder. It is up to the parents to decide if they are willing and able to have a child that will require special commitment or care.	**Reduces suffering of victims and families** Genetic screening gives couples **a choice** of whether to allow the birth of a child with a genetic defect. It is unethical to bring a child with a severe genetic disease into the world if this will result in substantial suffering of the individual, greatly reduce the happiness of parents and family, or drain the financial resources of society.
Uncertainty The reliability of tests may be in question: A false, positive or negative, result may cause parents to abort **healthy** embryos.	**Allows individuals to make lifestyle changes to deal with the disorder** If genetic screening reveals that a person has a genetic predisposition for cardio vascular diseases (CVD) then the person could make lifestyle changes to reduce the risk of developing CVD. Likewise, if a couple know that they are getting a thalassaemic child, they could make psychological and financial preparations to cope with the upbringing of the child.
Discrimination – a social issue Many diseases we can test for have no treatments. This could provide a basis for employers, insurance companies, banks, etc. to discriminate people with these disorders.	
Moral issues Moral issues are associated with embryo selection, because many unused embryos will be eventually discarded, especially in pre-implantation genetic diagnosis.	

Identification of genetic carriers (Heterozygotes) by DNA profiling

1. **Isolate and purify DNA** from embryonic cells (to test for preimplantation genetic disorders) or adult cells to identify heterozygotes. Extraction is normally carried out by mixing a tissue sample with a solvent, which dissolves DNA but precipitates protein (for example, a mixture of water-saturated phenol and chloroform). **Polymerase chain reaction** can be carried out to amplify (increase) the quantity of the DNA sample, if the sample is too small.

2. **Cut the DNA into fragments of variable length by using restriction endonucleases**: The enzymes cut the DNA at specific points called as recognition sites. The resulting fragments are called as restriction fragments. (*Refer to notes on restriction endonucleases*)

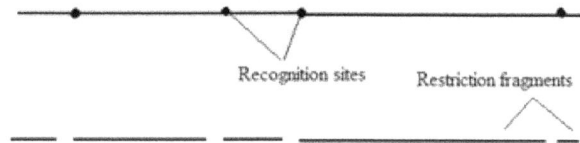

Recognition sites   Restriction fragments

Fig.22.8

3. **Separate the fragments by using gel electrophoresis.**
   - The fragments of DNA are placed into a well on an agarose gel plate.
   - The gel is flooded with an aqueous alkali at pH 8. The alkali is a good conductor of electricity.
   - Two electrodes (with a potential difference of 10mV) are immersed into the alkali on either side of the plate.
   - All the DNA fragments move towards the positive electrode because they have negatively charged phosphate groups at pH 8.
   - The smaller fragments move faster through the gel than the larger ones.
   - Thus, gel electrophoresis distributes fragments on to the gel according to their size.

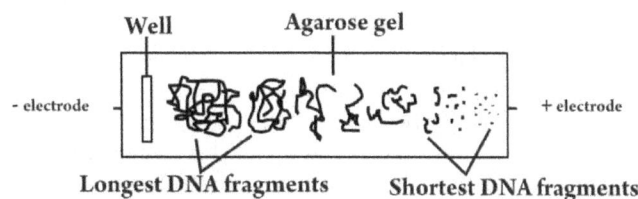

Note: DNA fragments cannot be seen.
The diagram is only an illustration.

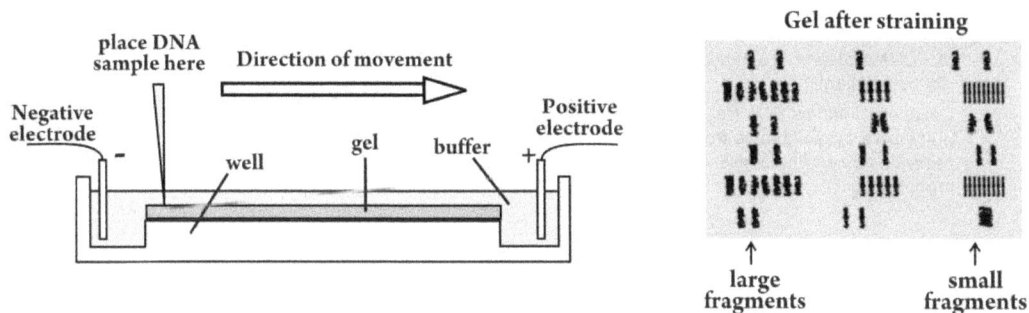

Fig.22.9

Unfortunately the DNA on the gel cannot be seen, so it must be visualised. There are three common methods for doing this: The gel can be stained with a chemical that specifically stains DNA, such as ethidium bromide or azure A. The DNA shows up as blue bands. This method is simple but not very sensitive. The DNA samples at the beginning can be radio labelled with a radioactive isotope such as ^{32}P. Photographic film is placed on top of the finished gel in the dark, and the DNA shows up as dark bands on the film. This method is extremely sensitive. The DNA fragments at the beginning can be labelled with a fluorescent molecule. The DNA fragments show up as coloured lights when the finished gel is illuminated with invisible ultraviolet light.

4. **Opening up the two DNA strands:** Up until this point, the DNA is double stranded. After electrophoresis, it is converted into single strands by immersing the gel in an alkali. This step is necessary to enable hybridization of the probe with the restriction fragments.

5. **Transfer DNA fragments from gel on to a nylon sheet by Southern blotting**
   A thin nylon membrane is laid over the gel and several sheets of blotting paper or filter paper are laid on top. The buffer containing the DNA is drawn up through the filter paper by capillarity and some of the DNA is deposited on the nylon membrane; the positions of the DNA fragments on the membrane correspond with their positions on the gel.

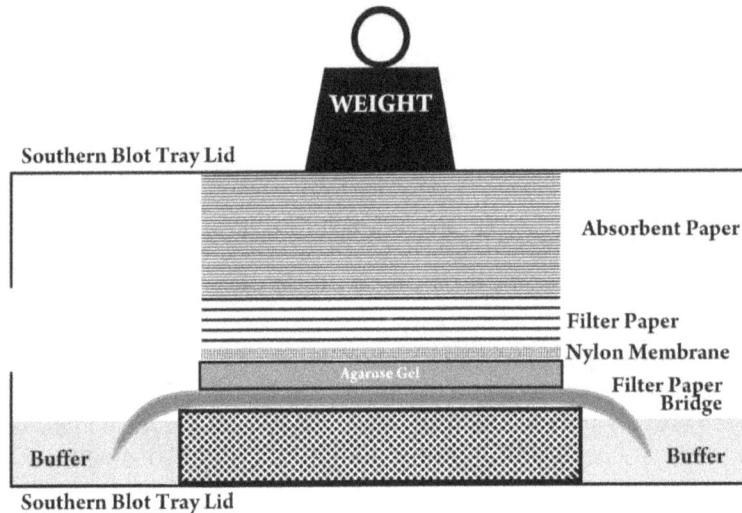

Fig.22.10

6. **Fixing Fragments to the membrane.** The DNA is fixed on the membrane by exposing it to short-wavelength ultraviolet light. This prevents the fragments from being washed away.

7. **Hybridisation of probes with complementary DNA on the nylon sheet**
   The separated single-stranded DNA is mixed with probes. Probes are always single-stranded, and can be made of DNA or RNA.

The probes contain nucleotide sequences that are complementary to the gene sequence that is being sought (e.g. the mutant gene which causes cystic fibrosis). The probes are labelled with substances that will cause black bands on the X-ray film. After hybridization (binding of probes with complementary base sequences on the DNA), the membrane is placed on an X-ray film in the dark. This causes the film to show dark bands where the probes are bound to the DNA (Mutant disease causing sequences). Black bands will form only if the mutant allele is present. If the normal allele alone is present, there will be **no black band at the domain of the defective gene**.

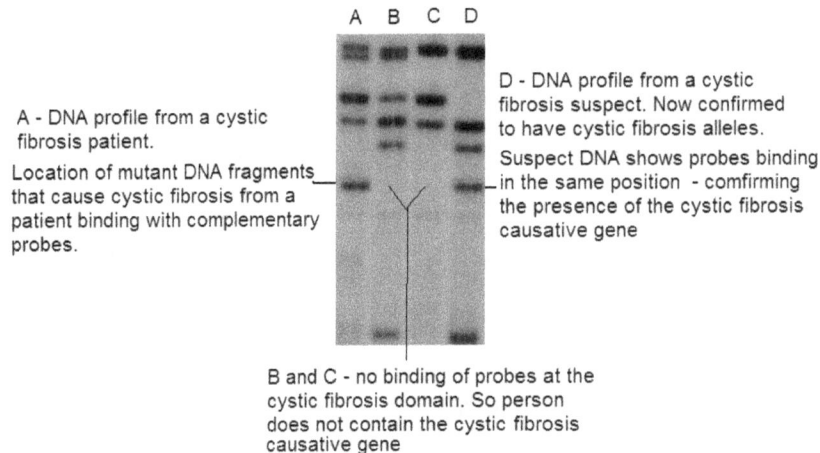

A - DNA profile from a cystic fibrosis patient.

Location of mutant DNA fragments that cause cystic fibrosis from a patient binding with complementary probes.

D - DNA profile from a cystic fibrosis suspect. Now confirmed to have cystic fibrosis alleles.

Suspect DNA shows probes binding in the same position - comfirming the presence of the cystic fibrosis causative gene

B and C - no binding of probes at the cystic fibrosis domain. So person does not contain the cystic fibrosis causative gene

Fig.22.11

Alternate method (DNA Microarray) – Additional information
To identify genetic defects. DNA probes have been prepared that match the sequences of many human genetic disease genes such as thalassaemia and cystic fibrosis. Hundreds of these probes can be stuck to a glass slide in a grid pattern, forming a _DNA microarray_ (or _DNA chip_). A sample of human DNA is added to the array and any sequences that match any of the various probes will stick to the array and be labelled. This allows rapid testing for a large number of genetic defects at a time.

………… **All the best for your exams** ………..

www.ingramcontent.com/pod-product-compliance
Lightning Source LLC
Chambersburg PA
CBHW081154090426

42736CB00017B/3314